FREE Test Taking Tips DVD Offer

To help us better serve you, we have developed a Test Taking Tips DVD that we would like to give you for FREE. **This DVD covers world-class test taking tips that you can use to be even more successful when you are taking your test.**

All that we ask is that you email us your feedback about your study guide. Please let us know what you thought about it – whether that is good, bad or indifferent.

To get your **FREE Test Taking Tips DVD**, email freedvd@studyguideteam.com with "FREE DVD" in the subject line and the following information in the body of the email:

    a. The title of your study guide.

    b. Your product rating on a scale of 1-5, with 5 being the highest rating.

    c. Your feedback about the study guide. What did you think of it?

    d. Your full name and shipping address to send your free DVD.

If you have any questions or concerns, please don't hesitate to contact us at freedvd@studyguideteam.com.

Thanks again!

PSAT 8/9 Math Workbook
PSAT 8/9 Math Prep with 2 Practice Tests [2nd Edition]

Test Prep Books

Copyright © 2020 by Test Prep Books

All rights reserved. No part of this publication may be reproduced, distributed, or transmitted in any form or by any means, including photocopying, recording, or other electronic or mechanical methods, without the prior written permission of the publisher, except in the case of brief quotations embodied in critical reviews and certain other noncommercial uses permitted by copyright law.

Written and edited by Test Prep Books.

Test Prep Books is not associated with or endorsed by any official testing organization. Test Prep Books is a publisher of unofficial educational products. All test and organization names are trademarks of their respective owners. Content in this book is included for utilitarian purposes only and does not constitute an endorsement by Test Prep Books of any particular point of view.

Interested in buying more than 10 copies of our product? Contact us about bulk discounts:
bulkorders@studyguideteam.com

ISBN 13: 9781628457605
ISBN 10: 1628457600

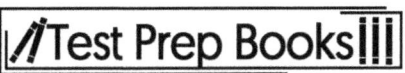

Table of Contents

Quick Overview *1*

Test-Taking Strategies *2*

FREE DVD OFFER *6*

Introduction to the PSAT 8/9 .... *7*

Math Test *9*
    Heart of Algebra 9
    Problem-Solving and Data Analysis 22
    Passport to Advanced Math 34

PSAT 8/9 Math Practice Test #1 .... *54*

Answer Explanations for Practice Test #1 .... *65*

PSAT 8/9 Math Practice Test #2 .... *75*

Answer Explanations for Practice Test #2 .... *85*

Quick Overview

As you draw closer to taking your exam, effective preparation becomes more and more important. Thankfully, you have this study guide to help you get ready. Use this guide to help keep your studying on track and refer to it often.

This study guide contains several key sections that will help you be successful on your exam. The guide contains tips for what you should do the night before and the day of the test. Also included are test-taking tips. Knowing the right information is not always enough. Many well-prepared test takers struggle with exams. These tips will help equip you to accurately read, assess, and answer test questions.

A large part of the guide is devoted to showing you what content to expect on the exam and to helping you better understand that content. In this guide are practice test questions so that you can see how well you have grasped the content. Then, answer explanations are provided so that you can understand why you missed certain questions.

Don't try to cram the night before you take your exam. This is not a wise strategy for a few reasons. First, your retention of the information will be low. Your time would be better used by reviewing information you already know rather than trying to learn a lot of new information. Second, you will likely become stressed as you try to gain a large amount of knowledge in a short amount of time. Third, you will be depriving yourself of sleep. So be sure to go to bed at a reasonable time the night before. Being well-rested helps you focus and remain calm.

Be sure to eat a substantial breakfast the morning of the exam. If you are taking the exam in the afternoon, be sure to have a good lunch as well. Being hungry is distracting and can make it difficult to focus. You have hopefully spent lots of time preparing for the exam. Don't let an empty stomach get in the way of success!

When travelling to the testing center, leave earlier than needed. That way, you have a buffer in case you experience any delays. This will help you remain calm and will keep you from missing your appointment time at the testing center.

Be sure to pace yourself during the exam. Don't try to rush through the exam. There is no need to risk performing poorly on the exam just so you can leave the testing center early. Allow yourself to use all of the allotted time if needed.

Remain positive while taking the exam even if you feel like you are performing poorly. Thinking about the content you should have mastered will not help you perform better on the exam.

Once the exam is complete, take some time to relax. Even if you feel that you need to take the exam again, you will be well served by some down time before you begin studying again. It's often easier to convince yourself to study if you know that it will come with a reward!

Test-Taking Strategies

1. Predicting the Answer

When you feel confident in your preparation for a multiple-choice test, try predicting the answer before reading the answer choices. This is especially useful on questions that test objective factual knowledge. By predicting the answer before reading the available choices, you eliminate the possibility that you will be distracted or led astray by an incorrect answer choice. You will feel more confident in your selection if you read the question, predict the answer, and then find your prediction among the answer choices. After using this strategy, be sure to still read all of the answer choices carefully and completely. If you feel unprepared, you should not attempt to predict the answers. This would be a waste of time and an opportunity for your mind to wander in the wrong direction.

2. Reading the Whole Question

Too often, test takers scan a multiple-choice question, recognize a few familiar words, and immediately jump to the answer choices. Test authors are aware of this common impatience, and they will sometimes prey upon it. For instance, a test author might subtly turn the question into a negative, or he or she might redirect the focus of the question right at the end. The only way to avoid falling into these traps is to read the entirety of the question carefully before reading the answer choices.

3. Looking for Wrong Answers

Long and complicated multiple-choice questions can be intimidating. One way to simplify a difficult multiple-choice question is to eliminate all of the answer choices that are clearly wrong. In most sets of answers, there will be at least one selection that can be dismissed right away. If the test is administered on paper, the test taker could draw a line through it to indicate that it may be ignored; otherwise, the test taker will have to perform this operation mentally or on scratch paper. In either case, once the obviously incorrect answers have been eliminated, the remaining choices may be considered. Sometimes identifying the clearly wrong answers will give the test taker some information about the correct answer. For instance, if one of the remaining answer choices is a direct opposite of one of the eliminated answer choices, it may well be the correct answer. The opposite of obviously wrong is obviously right! Of course, this is not always the case. Some answers are obviously incorrect simply because they are irrelevant to the question being asked. Still, identifying and eliminating some incorrect answer choices is a good way to simplify a multiple-choice question.

4. Don't Overanalyze

Anxious test takers often overanalyze questions. When you are nervous, your brain will often run wild, causing you to make associations and discover clues that don't actually exist. If you feel that this may be a problem for you, do whatever you can to slow down during the test. Try taking a deep breath or counting to ten. As you read and consider the question, restrict yourself to the particular words used by the author. Avoid thought tangents about what the author *really* meant, or what he or she was *trying* to say. The only things that matter on a multiple-choice test are the words that are actually in the question. You must avoid reading too much into a multiple-choice question, or supposing that the writer meant something other than what he or she wrote.

5. No Need for Panic

It is wise to learn as many strategies as possible before taking a multiple-choice test, but it is likely that you will come across a few questions for which you simply don't know the answer. In this situation, avoid panicking. Because most multiple-choice tests include dozens of questions, the relative value of a single wrong answer is small. As much as possible, you should compartmentalize each question on a multiple-choice test. In other words, you should not allow your feelings about one question to affect your success on the others. When you find a question that you either don't understand or don't know how to answer, just take a deep breath and do your best. Read the entire question slowly and carefully. Try rephrasing the question a couple of different ways. Then, read all of the answer choices carefully. After eliminating obviously wrong answers, make a selection and move on to the next question.

6. Confusing Answer Choices

When working on a difficult multiple-choice question, there may be a tendency to focus on the answer choices that are the easiest to understand. Many people, whether consciously or not, gravitate to the answer choices that require the least concentration, knowledge, and memory. This is a mistake. When you come across an answer choice that is confusing, you should give it extra attention. A question might be confusing because you do not know the subject matter to which it refers. If this is the case, don't eliminate the answer before you have affirmatively settled on another. When you come across an answer choice of this type, set it aside as you look at the remaining choices. If you can confidently assert that one of the other choices is correct, you can leave the confusing answer aside. Otherwise, you will need to take a moment to try to better understand the confusing answer choice. Rephrasing is one way to tease out the sense of a confusing answer choice.

7. Your First Instinct

Many people struggle with multiple-choice tests because they overthink the questions. If you have studied sufficiently for the test, you should be prepared to trust your first instinct once you have carefully and completely read the question and all of the answer choices. There is a great deal of research suggesting that the mind can come to the correct conclusion very quickly once it has obtained all of the relevant information. At times, it may seem to you as if your intuition is working faster even than your reasoning mind. This may in fact be true. The knowledge you obtain while studying may be retrieved from your subconscious before you have a chance to work out the associations that support it. Verify your instinct by working out the reasons that it should be trusted.

8. Key Words

Many test takers struggle with multiple-choice questions because they have poor reading comprehension skills. Quickly reading and understanding a multiple-choice question requires a mixture of skill and experience. To help with this, try jotting down a few key words and phrases on a piece of scrap paper. Doing this concentrates the process of reading and forces the mind to weigh the relative importance of the question's parts. In selecting words and phrases to write down, the test taker thinks about the question more deeply and carefully. This is especially true for multiple-choice questions that are preceded by a long prompt.

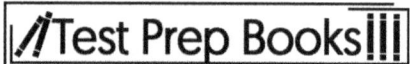

9. Subtle Negatives

One of the oldest tricks in the multiple-choice test writer's book is to subtly reverse the meaning of a question with a word like *not* or *except*. If you are not paying attention to each word in the question, you can easily be led astray by this trick. For instance, a common question format is, "Which of the following is…?" Obviously, if the question instead is, "Which of the following is not…?," then the answer will be quite different. Even worse, the test makers are aware of the potential for this mistake and will include one answer choice that would be correct if the question were not negated or reversed. A test taker who misses the reversal will find what he or she believes to be a correct answer and will be so confident that he or she will fail to reread the question and discover the original error. The only way to avoid this is to practice a wide variety of multiple-choice questions and to pay close attention to each and every word.

10. Reading Every Answer Choice

It may seem obvious, but you should always read every one of the answer choices! Too many test takers fall into the habit of scanning the question and assuming that they understand the question because they recognize a few key words. From there, they pick the first answer choice that answers the question they believe they have read. Test takers who read all of the answer choices might discover that one of the latter answer choices is actually *more* correct. Moreover, reading all of the answer choices can remind you of facts related to the question that can help you arrive at the correct answer. Sometimes, a misstatement or incorrect detail in one of the latter answer choices will trigger your memory of the subject and will enable you to find the right answer. Failing to read all of the answer choices is like not reading all of the items on a restaurant menu: you might miss out on the perfect choice.

11. Spot the Hedges

One of the keys to success on multiple-choice tests is paying close attention to every word. This is never truer than with words like almost, most, some, and sometimes. These words are called "hedges" because they indicate that a statement is not totally true or not true in every place and time. An absolute statement will contain no hedges, but in many subjects, the answers are not always straightforward or absolute. There are always exceptions to the rules in these subjects. For this reason, you should favor those multiple-choice questions that contain hedging language. The presence of qualifying words indicates that the author is taking special care with his or her words, which is certainly important when composing the right answer. After all, there are many ways to be wrong, but there is only one way to be right! For this reason, it is wise to avoid answers that are absolute when taking a multiple-choice test. An absolute answer is one that says things are either all one way or all another. They often include words like *every*, *always*, *best*, and *never*. If you are taking a multiple-choice test in a subject that doesn't lend itself to absolute answers, be on your guard if you see any of these words.

12. Long Answers

In many subject areas, the answers are not simple. As already mentioned, the right answer often requires hedges. Another common feature of the answers to a complex or subjective question are qualifying clauses, which are groups of words that subtly modify the meaning of the sentence. If the question or answer choice describes a rule to which there are exceptions or the subject matter is complicated, ambiguous, or confusing, the correct answer will require many words in order to be expressed clearly and accurately. In essence, you should not be deterred by answer choices that seem excessively long. Oftentimes, the author of the text will not be able to write the correct answer without offering some qualifications and modifications. Your job is to read the answer choices thoroughly and completely and to select the one that most accurately and precisely answers the question.

13. Restating to Understand

Sometimes, a question on a multiple-choice test is difficult not because of what it asks but because of how it is written. If this is the case, restate the question or answer choice in different words. This process serves a couple of important purposes. First, it forces you to concentrate on the core of the question. In order to rephrase the question accurately, you have to understand it well. Rephrasing the question will concentrate your mind on the key words and ideas. Second, it will present the information to your mind in a fresh way. This process may trigger your memory and render some useful scrap of information picked up while studying.

14. True Statements

Sometimes an answer choice will be true in itself, but it does not answer the question. This is one of the main reasons why it is essential to read the question carefully and completely before proceeding to the answer choices. Too often, test takers skip ahead to the answer choices and look for true statements. Having found one of these, they are content to select it without reference to the question above. Obviously, this provides an easy way for test makers to play tricks. The savvy test taker will always read the entire question before turning to the answer choices. Then, having settled on a correct answer choice, he or she will refer to the original question and ensure that the selected answer is relevant. The mistake of choosing a correct-but-irrelevant answer choice is especially common on questions related to specific pieces of objective knowledge. A prepared test taker will have a wealth of factual knowledge at his or her disposal, and should not be careless in its application.

15. No Patterns

One of the more dangerous ideas that circulates about multiple-choice tests is that the correct answers tend to fall into patterns. These erroneous ideas range from a belief that B and C are the most common right answers, to the idea that an unprepared test-taker should answer "A-B-A-C-A-D-A-B-A." It cannot be emphasized enough that pattern-seeking of this type is exactly the WRONG way to approach a multiple-choice test. To begin with, it is highly unlikely that the test maker will plot the correct answers according to some predetermined pattern. The questions are scrambled and delivered in a random order. Furthermore, even if the test maker was following a pattern in the assignation of correct answers, there is no reason why the test taker would know which pattern he or she was using. Any attempt to discern a pattern in the answer choices is a waste of time and a distraction from the real work of taking the test. A test taker would be much better served by extra preparation before the test than by reliance on a pattern in the answers.

FREE DVD OFFER

Don't forget that doing well on your exam includes both understanding the test content and understanding how to use what you know to do well on the test. We offer a completely FREE Test Taking Tips DVD that covers world class test taking tips that you can use to be even more successful when you are taking your test.

All that we ask is that you email us your feedback about your study guide. To get your **FREE Test Taking Tips DVD**, email freedvd@studyguideteam.com with "FREE DVD" in the subject line and the following information in the body of the email:

- The title of your study guide.
- Your product rating on a scale of 1-5, with 5 being the highest rating.
- Your feedback about the study guide. What did you think of it?
- Your full name and shipping address to send your free DVD.

Introduction to the PSAT 8/9

Function of the Test

The Preliminary SAT (PSAT) 8/9 is an introductory version of the PSAT/NMSQT, PSAT 10, and ultimately, the SAT. Given by the College Board, the PSAT 8/9 is designed to help U.S. eighth- and ninth-grade students determine their areas of weakness so that they can focus their preparation for the future testing, and most importantly, optimize their success in college. Because it assesses the same knowledge and skills that the subsequent tests measure but in an understandable form for eighth and ninth graders, the results can serve as a strong personal benchmark for test takers in terms of their areas of competency and those that need the most improvement. As an added benefit, test takers over the age of thirteen can receive a free tailored practice experience, courtesy of Khan Academy® if they opt to submit their scores.

The PSAT 8/9 should be used as a tool to gauge mastery of concepts learned in school and progress towards those needed to succeed in college. The College Board recommends that the best way to prepare for the exam is simply to work hard in classes, complete all assignments, study regularly, remain inquisitive and active in class, and take advantage of challenging courses that are offered. While cramming for the exam is not necessary and not likely helpful, a favorite strategy of successful test takers and students is to familiarize themselves with the types of questions asked, the format of the questions, and the content that will be measured on the exam by taking practice exams and using study guides specifically created for the PSAT and SAT suite of tests.

Test Administration

The PSAT 8/9 is offered to students in the eighth and ninth grades on various dates during the academic year at schools throughout the United States. Some schools will pay all or part of the exam registration fee for their pupils. Since the financial responsibility of the student for the exam is different for each school, it is best to consult the school's guidance department for specifics. International and homeschooled students can also take the PSAT 8/9 by contacting local schools or using the School Search page on the College Board website to locate a site where the test will be administered.

Students with documented disabilities can contact the College Board to make alternative arrangements to take the PSAT 8/9. Examples of accommodations that are permitted include extending testing or break time and reading and seeing aids. College Board approval is not required to grant the accommodations, although nonstandard test format requests (such as a test book printed in Braille) must be received by the College Board before the ordering deadline.

On the test day, test takers should bring a couple of No. 2 pencils, an eraser, a calculator approved by the College Board, and a valid school- or government-issued ID. No other materials, except those approved for disability accommodations, are permitted.

Test Format

The PSAT 8/9 gauges a student's proficiency in three areas: Reading, Mathematics, and Writing and Language. All the tests that fall under the PSAT and SAT umbrella were redesigned in 2015. The PSAT 8/9 is very similar to the new PSAT and SAT in substance, structure, and scoring methodology, except that, like the PSAT, it does not include an essay.

The reading portion of the PSAT measures comprehension, requiring candidates to read multi-paragraph fiction and non-fiction segments including informational visuals, such as charts, tables and graphs, and answer questions based on this content. Three critical sectors are tested for the Math section: Solving problems and analyzing data, Algebra, and complex equations and operations. The writing and language portion requires students to evaluate and edit writing and graphics to obtain an answer that correctly conveys the information given in the passage.

The PSAT 8/9 contains 7 grid-in math questions, and 113 multiple-choice questions, with each section comprising around 40 questions. A different length of time is given for each section, for a total of two hours, 35 minutes. The specific breakdown is as follows:

Section	Time (In Minutes)	Number of Questions
Reading	55	42
Writing and Language	40	40
Mathematics	60 total 40 with calculator 20 without calculator	38 total 31 Multiple-Choice 7 Grid-in
Total	155	120

Scoring

Test takers are not penalized for incorrect answers on the PSAT 8/9, so it is smart to guess, even when the test taker does not know the answer. A raw school is calculated based on the number of correct responses. This score is then equated, or scaled, to a total score that ranges from 240 to 1440. Of this, 120 to 720 is contributed from each of two subsections: Math and Evidence-Based Reading and Writing. The score equating process allows different iterations and administrations of the PSAT 8/9 to be compared. Score reports also list sub-scores for math, reading, and writing on a scale from 6 to 36, in order to give candidates an idea of their specific strengths and weaknesses. Mean, or average, scores received by characteristic U.S. test takers, are broken down by grade level.

The report ranks scores based on a percentile between 1 and 99 so students can see how they measured up to other test takers. The average (50th percentile) for each subtest is listed as well. Good scores are typically defined as higher than 50 percent. Benchmark scores are also provided. These increase by grade level and serve to indicate whether the candidate is on track for success in college based on his or her relative achievement on the benchmark continuum. For eighth grade students, the benchmark for the Math portion is a score of 430, while a score of 390 is set as the benchmark on the Evidence-Based Reading and Writing subtest. These scores each increase by 20 points for ninth grade test takers.

Scores are not sent to colleges. The College Board only sends PSAT 8/9 scores to schools, and usually districts and states. A copy is also sent to parents by some schools directly.

Recent/Future Developments

In the coming years, instead of comparing a test taker's score range, average score, and percentile to the population of the previous year's test takers, these metrics will be compared to norm groups that will be derived from research data.

Math Test

Heart of Algebra

Creating, Solving, or Interpreting a Linear Expression or Equation in One Variable

Linear expressions and equations are concise mathematical statements that can be written to model a variety of scenarios. Questions found pertaining to this topic will contain one variable only. A variable is an unknown quantity, usually denoted by a letter (x, n, p, etc.). In the case of linear expressions and equations, the power of the variable (its exponent) is 1. A variable without a visible exponent is raised to the first power.

<u>Writing Linear Expressions and Equations</u>
A linear expression is a statement about an unknown quantity expressed in mathematical symbols. The statement "five times a number added to forty" can be expressed as $5x + 40$. A linear equation is a statement in which two expressions (at least one containing a variable) are equal to each other. The statement "five times a number added to forty is equal to ten" can be expressed as $5x + 40 = 10$. Real-world scenarios can also be expressed mathematically. Consider the following:

> Bob had $20 and Tom had $4. After selling 4 ice cream cones to Bob, Tom has as much money as Bob.

The cost of an ice cream cone is an unknown quantity and can be represented by a variable. The amount of money Bob has after his purchase is four times the cost of an ice cream cone subtracted from his original $20. The amount of money Tom has after his sale is four times the cost of an ice cream cone added to his original $4. This can be expressed as: $20 - 4x = 4x + 4$, where x represents the cost of an ice cream cone.

When expressing a verbal or written statement mathematically, it is key to understand words or phrases that can be represented with symbols. The following are examples:

Symbol	Phrase
+	added to, increased by, sum of, more than
−	decreased by, difference between, less than, take away
×	multiplied by, 3 (4, 5 . . .) times as large, product of
÷	divided by, quotient of, half (third, etc.) of
=	is, the same as, results in, as much as
$x, t, n, etc.$	a number, unknown quantity, value of

<u>Evaluating and Simplifying Algebraic Expressions</u>
Given an algebraic expression, students may be asked to evaluate for given values of variable(s). In doing so, students will arrive at a numerical value as an answer. For example:

$$\text{Evaluate } a - 2b + ab \text{ for } a = 3 \text{ and } b = -1$$

To evaluate an expression, the given values should be substituted for the variables and simplified using the order of operations. In this case: $(3) - 2(-1) + (3)(-1)$. Parentheses are used when substituting.

Given an algebraic expression, students may be asked to simplify the expression. For example:

$$\text{Simplify } 5x^2 - 10x + 2 - 8x^2 + x - 1.$$

Simplifying algebraic expressions requires combining like terms. A term is a number, variable, or product of a number and variables separated by addition and subtraction. The terms in the above expressions are: $5x^2, -10x, 2, -8x^2, x,$ and -1.

Like terms have the same variables raised to the same powers (exponents). To combine like terms, the coefficients (numerical factor of the term including sign) are added, while the variables and their powers are kept the same. The example above simplifies to $-3x^2 - 9x + 1$.

Solving Linear Equations

When asked to solve a linear equation, it requires determining a numerical value for the unknown variable. Given a linear equation involving addition, subtraction, multiplication, and division, isolation of the variable is done by working backward. Addition and subtraction are inverse operations, as are multiplication and division; therefore, they can be used to cancel each other out.

The first steps to solving linear equations are to distribute if necessary and combine any like terms that are on the same side of the equation. Sides of an equation are separated by an = sign. Next, the equation should be manipulated to get the variable on one side. Whatever is done to one side of an equation, must be done to the other side to remain equal. Then, the variable should be isolated by using inverse operations to undo the order of operations backward. Undo addition and subtraction, then undo multiplication and division. For example:

Solve $4(t - 2) + 2t - 4 = 2(9 - 2t)$

Distribute: $4t - 8 + 2t - 4 = 18 - 4t$

Combine like terms: $6t - 12 = 18 - 4t$

Add 4t to each side to move the variable: $10t - 12 = 18$

Add 12 to each side to isolate the variable: $10t = 30$

Divide each side by 10 to isolate the variable: $t = 3$

The answer can be checked by substituting the value for the variable into the original equation and ensuring both sides calculate to be equal.

Creating, Solving, or Interpreting Linear Inequalities in One Variable

Linear inequalities and linear equations are both comparisons of two algebraic expressions. However, unlike equations in which the expressions are equal to each other, linear inequalities compare expressions that are unequal. Linear equations typically have one value for the variable that makes the statement true. Linear inequalities generally have an infinite number of values that make the statement true.

Writing Linear Inequalities

Linear inequalities are a concise mathematical way to express the relationship between unequal values. More specifically, they describe in what way the values are unequal. A value could be greater than ($>$); less than ($<$); greater than or equal to ($\geq$); or less than or equal to ($\leq$) another value. The statement "five times a number added to forty is more than sixty-five" can be expressed as $5x + 40 > 65$. Common words and phrases that express inequalities are:

Symbol	Phrase
$<$	is under, is below, smaller than, beneath
$>$	is above, is over, bigger than, exceeds
$\leq$	no more than, at most, maximum
$\geq$	no less than, at least, minimum

Solving Linear Inequalities

When solving a linear inequality, the solution is the set of all numbers that makes the statement true. The inequality $x + 2 \geq 6$ has a solution set of 4 and every number greater than 4 (4.0001, 5, 12, 107, etc.). Adding 2 to 4 or any number greater than 4 would result in a value that is greater than or equal to 6. Therefore, $x \geq 4$ would be the solution set.

Solution sets for linear inequalities often will be displayed using a number line. If a value is included in the set ($\geq$ or $\leq$), there is a shaded dot placed on that value and an arrow extending in the direction of the solutions. For a variable $>$ or $\geq$ a number, the arrow would point right on the number line (the direction where the numbers increase); and if a variable is $<$ or $\leq$ a number, the arrow would point left (where the numbers decrease). If the value is not included in the set ($>$ or $<$), an open circle on that value would be used with an arrow in the appropriate direction.

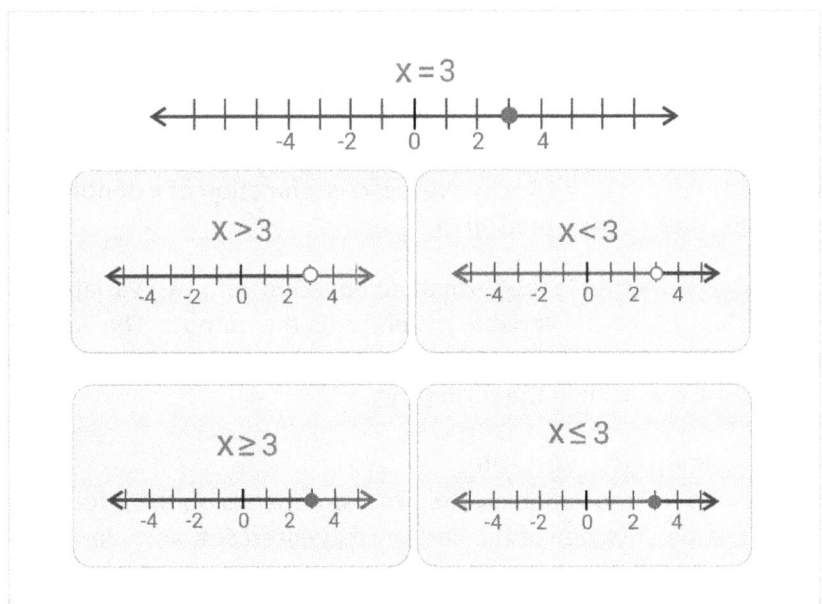

Students may be asked to write a linear inequality given a graph of its solution set. To do so, they should identify whether the value is included (shaded dot or open circle) and the direction in which the arrow is pointing.

In order to algebraically solve a linear inequality, the same steps should be followed as in solving a linear equation. The inequality symbol stays the same for all operations EXCEPT when dividing by a negative number. If dividing by a negative number while solving an inequality, the relationship reverses (the sign flips). Dividing by a positive does not change the relationship, so the sign stays the same. In other words, > switches to < and vice versa. An example is shown below.

Solve $-2(x + 4) \leq 22$

Distribute: $-2x - 8 \leq 22$

Add 8 to both sides: $-2x \leq 30$

Divide both sides by -2: $x \geq 15$

Building a Linear Function that Models a Linear Relationship Between Two Quantities

Linear relationships between two quantities can be expressed in two ways: function notation or as a linear equation with two variables. The relationship is referred to as linear because its graph is represented by a line. For a relationship to be linear, both variables must be raised to the first power only.

Function/Linear Equation Notation

A relation is a set of input and output values that can be written as ordered pairs. A function is a relation in which each input is paired with exactly one output. The domain of a function consists of all inputs, and the range consists of all outputs. Graphing the ordered pairs of a linear function produces a straight line. An example of a function would be $f(x) = 4x + 4$, read "f of x is equal to four times x plus four."

In this example, the input would be x and the output would be f(x). Ordered pairs would be represented as (x, f(x)). To find the output for an input value of 3, 3 would be substituted for x into the function as follows: $f(3) = 4(3) + 4$, resulting in $f(3) = 16$.

Therefore, the ordered pair $(3, f(3)) = (3, 16)$. Note f(x) is a function of x denoted by f. Functions of x could be named g(x), read "g of x"; p(x), read "p of x"; etc.

A linear function could also be written in the form of an equation with two variables. Typically, the variable x represents the inputs and the variable y represents the outputs. The variable x is considered the independent variable and y the dependent variable. The above function would be written as $y = 4x + 4$. Ordered pairs are written in the form (x, y).

Writing Linear Equations in Two Variables

When writing linear equations in two variables, the process depends on the information given. Questions will typically provide the slope of the line and its y-intercept, an ordered pair and the slope, or two ordered pairs.

Given the Slope and Y-Intercept

Linear equations are commonly written in slope-intercept form, $y = mx + b$, where m represents the slope of the line and b represents the y-intercept. The slope is the rate of change between the variables, usually expressed as a whole number or fraction. The y-intercept is the value of y when x = 0 (the point where the line intercepts the y-axis on a graph). Given the slope and y-intercept of a line, the values are substituted for *m* and *b* into the equation. A line with a slope of ½ and *y*-intercept of -2 would have an equation $y = ½x - 2$.

Given an Ordered Pair and the Slope

The point-slope form of a line, $y - y_1 = m(x - x_1)$, is used to write an equation when given an ordered pair (point on the equation's graph) for the function and its rate of change (slope of the line). The values for the slope, *m*, and the point (x_1, y_1) are substituted into the point-slope form to obtain the equation of the line. A line with a slope of 3 and an ordered pair (4, -2) would have an equation $y - (-2) = 3(x - 4)$. If a question specifies that the equation be written in slope-intercept form, the equation should be manipulated to isolate *y*:

Solve: $y - (-2) = 3(x - 4)$

Distribute: $y + 2 = 3x - 12$

Subtract 2 from both sides: $y = 3x - 14$

Given Two Ordered Pairs

Given two ordered pairs for a function, (x_1, y_1) and (x_2, y_2), it is possible to determine the rate of change between the variables (slope of the line). To calculate the slope of the line, m, the values for the ordered pairs should be substituted into the formula: $m = \frac{y_2 - y_1}{x_2 - x_1}$. The expression is substituted to obtain a whole number or fraction for the slope. Once the slope is calculated, the slope and either of the ordered pairs should be substituted into the point-slope form to obtain the equation of the line.

Creating, Solving, and Interpreting Systems of Linear Inequalities in Two Variables

Expressing Linear Inequalities in Two Variables

A linear inequality in two variables is a statement expressing an unequal relationship between those two variables. Typically written in slope-intercept form, the variable *y* can be greater than; less than; greater than or equal to; or less than or equal to a linear expression including the variable *x*. Examples include $y > 3x$ and $y \leq ½x - 3$. Questions may instruct students to model real world scenarios such as:

> You work part-time cutting lawns for $15 each and cleaning houses for $25 each. Your goal is to make more than $90 this week. Write an inequality to represent the possible pairs of lawns and houses needed to reach your goal.

This scenario can be expressed as $15x + 25y > 90$ where *x* is the number of lawns cut and *y* is the number of houses cleaned.

Graphing Solution Sets for Linear Inequalities in Two Variables

A graph of the solution set for a linear inequality shows the ordered pairs that make the statement true. The graph consists of a boundary line dividing the coordinate plane and shading on one side of the boundary. The boundary line should be graphed just as a linear equation would be graphed. If the inequality symbol is > or <, a dashed line can be used to indicate that the line is not part of the solution set. If the inequality symbol is ≥ or ≤, a solid line can be used to indicate that the boundary line is included in the solution set. An ordered pair (*x*, *y*) on either side of the line should be chosen to test in the inequality statement. If substituting the values for *x* and *y* results in a true statement $(15(3) + 25(2) > 90)$, that ordered pair and all others on that side of the boundary line are part of the solution set. To indicate this, that region of the graph should be shaded. If substituting the ordered pair results in a false statement, the ordered pair and all others on that side are not part of the solution set.

Therefore, the other region of the graph contains the solutions and should be shaded.

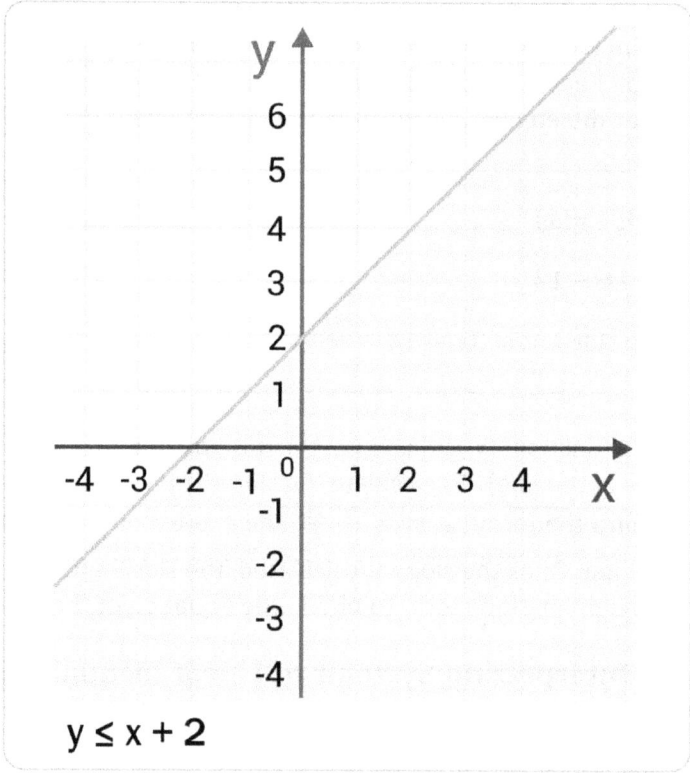

$y \leq x + 2$

A question may simply ask whether a given ordered pair is a solution to a given inequality. To determine this, the values should be substituted for the ordered pair into the inequality. If the result is a true statement, the ordered pair is a solution; if the result is a false statement, the ordered pair is not a solution.

Expressing Systems of Linear Inequalities in Two Variables
A system of linear inequalities consists of two linear inequalities making comparisons between two variables. Students may be given a scenario and asked to express it as a system of inequalities:

> A consumer study calls for at least 60 adult participants. It cannot use more than 25 men. Express these constraints as a system of inequalities.

This can be modeled by the system: $x + y \geq 60; x \leq 25$, where x represents the number of men and y represents the number of women. A solution to the system is an ordered pair that makes both inequalities true when substituting the values for x and y.

Graphing Solution Sets for Systems of Linear Inequalities in Two Variables
The solution set for a system of inequalities is the region of a graph consisting of ordered pairs that make both inequalities true. To graph the solution set, each linear inequality should first be graphed

with appropriate shading. The region of the graph should be identified where the shading for the two inequalities overlaps. This region contains the solution set for the system.

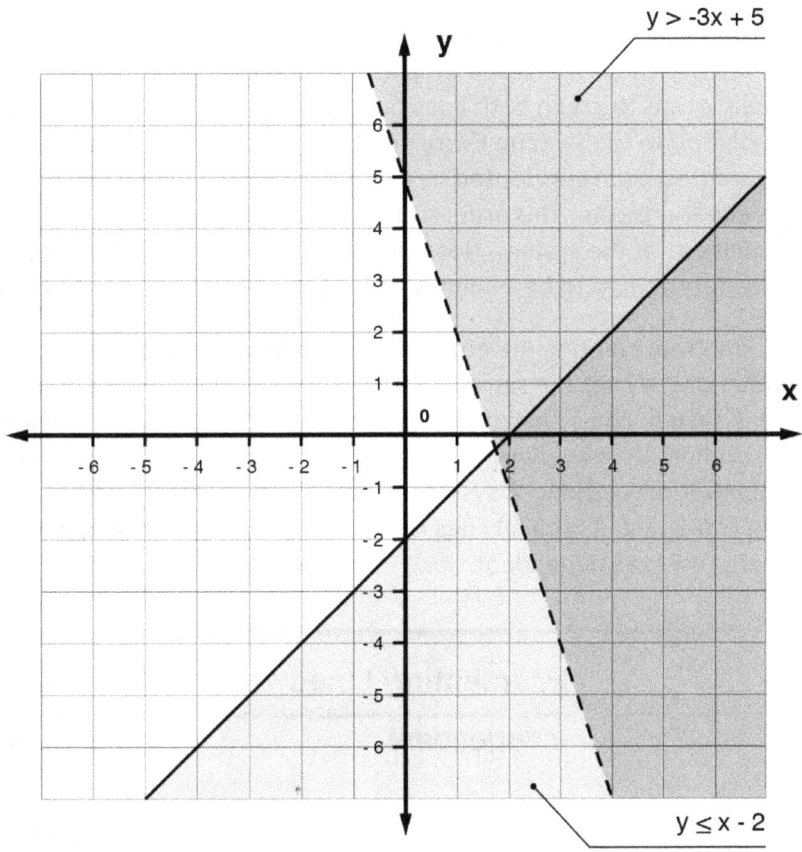

An ordered pair from the region of solutions can be selected to test in the system of inequalities.

Just as with manipulating linear inequalities in one variable, if dividing by a negative number in working with a linear inequality in two variables, the relationship reverses and the inequality sign should be flipped.

Creating, Solving, and Interpreting Systems of Two Linear Equations in Two Variables

Expressing Systems of Two Linear Equations in Two Variables

A system of two linear equations in two variables is a set of equations that use the same variables, usually x and y. Here's a sample problem:

> An Internet provider charges an installation fee and a monthly charge. It advertises that two months of its offering costs $100 and six months costs $200. Find the monthly charge and the installation fee.

The two unknown quantities (variables) are the monthly charge and the installation fee. There are two different statements given relating the variables: two months added to the installation fee is $100; and six months added to the installation fee is $200. Using the variable x as the monthly charge and y as the installation fee, the statements can be written as the following: $2x + y = 100; 6x + y = 200$. These two equations taken together form a system modeling the given scenario.

Solutions of a System of Two Linear Equations in Two Variables

A solution for a system of equations is an ordered pair that makes both equations true. One method for solving a system of equations is to graph both lines on a coordinate plane. If the lines intersect, the point of intersection is the solution to the system. Every point on a line represents an ordered pair that makes its equation true. The ordered pair represented by this point of intersection lies on both lines and therefore makes both equations true. This ordered pair should be checked by substituting its values into both of the original equations of the system. Note that given a system of equations and an ordered pair, the ordered pair can be determined to be a solution or not by checking it in both equations.

If, when graphed, the lines representing the equations of a system do not intersect, then the two lines are parallel to each other or they are the same exact line. Parallel lines extend in the same direction without ever meeting. A system consisting of parallel lines has no solution. If the equations for a system represent the same exact line, then every point on the line is a solution to the system. In this case, there would be an infinite number of solutions. A system consisting of intersecting lines is referred to as independent; a system consisting of parallel lines is referred to as inconsistent; and a system consisting of coinciding lines is referred to as dependent.

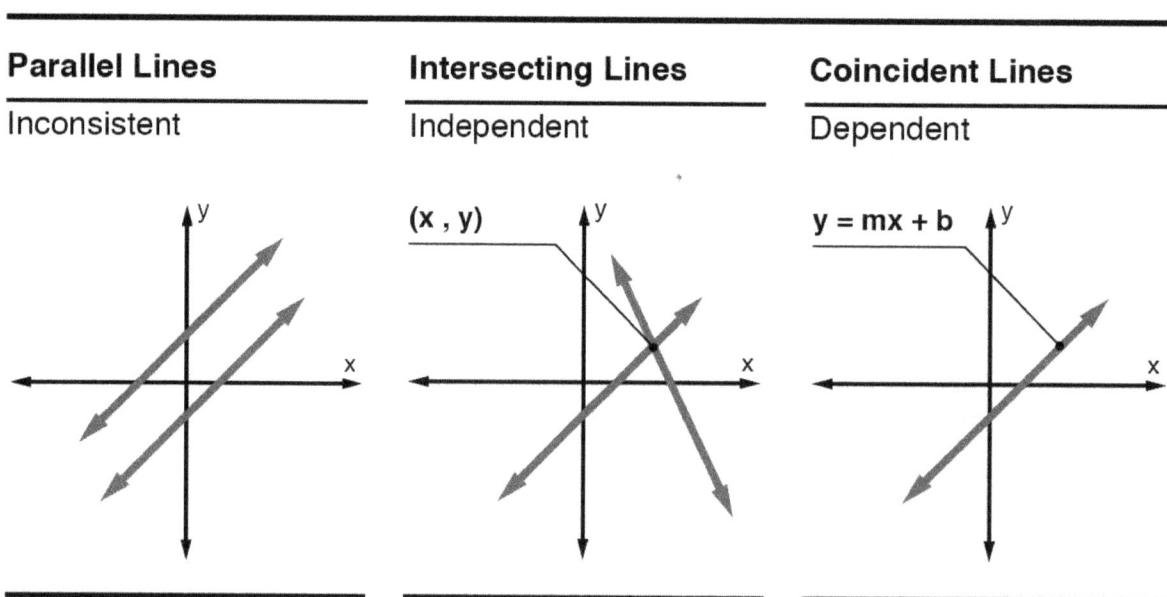

Algebraically Solving Linear Equations (or Inequalities) in One Variable

Linear equations in one variable and linear inequalities in one variable can be solved following similar processes. Although they typically have one solution, a linear equation can have no solution or can have a solution set of all real numbers. Solution sets for linear inequalities typically consist of an infinite number of values either greater or less than a given value (where the given value may or may not be included in the set). However, a linear inequality can have no solution or can have a solution set consisting of all real numbers.

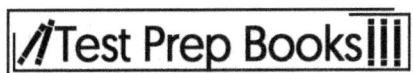

Linear Equations in One Variable – Special Cases

Solving a linear equation produces a value for the variable that makes the algebraic statement true. If there is no value for the variable that would make the statement true, there is no solution to the equation. Here's a sample equation: $x + 3 = x - 1$.

There is no value for *x* in which adding 3 to the value would produce the same result as subtracting 1 from that value. Conversely, if any value for the variable would make a true statement, the equation has an infinite number of solutions.

Here's another sample equation: $3x + 6 = 3(x + 2)$. Any real number substituted for *x* would result in a true statement (both sides of the equation are equal).

By manipulating equations similar to the two above, the variable of the equation will cancel out completely. If the constants that are left express a true statement (ex., $6 = 6$), then all real numbers are solutions to the equation. If the constants left express a false statement (ex., $3 = -1$), then there is no solution to the equation.

A question on this material may present a linear equation with an unknown value for either a constant or a coefficient of the variable and ask to determine the value that produces an equation with no solution or infinite solutions. For example:

$3x + 7 = 3x + 10 + n$; Find the value of *n* that would create an equation with an infinite number of solutions for the variable *x*.

To solve this problem, the equation should be manipulated so the variable *x* will cancel. To do this, $3x$ should be subtracted from both sides, which would leave $7 = 10 + n$. By subtracting 10 on both sides, it is determined that $n = -3$. Therefore, a value of -3 for *n* would result in an equation with a solution set of all real numbers.

If the same problem asked for the equation to have no solution, the value of *n* would be all real numbers except -3.

Linear Inequalities in One Variable – Special Cases

A linear inequality can have a solution set consisting of all real numbers or can contain no solution. When solved algebraically, a linear inequality in which the variable cancels out and results in a true statement (ex., $7 \geq 2$) has a solution set of all real numbers. A linear inequality in which the variable cancels out and results in a false statement (ex., $7 \leq 2$) has no solution.

Compound Inequalities

A compound inequality is a pair of inequalities joined by *and* or *or*. Given a compound inequality, to determine its solution set, both inequalities should be solved for the given variable. The solution set for a compound inequality containing *and* consists of all the values for the variable that make both inequalities true. If solving the compound inequality results in $x > -9$ and $x \leq 6$, the solution set would consist of all values between -9 and 6, including 6. This may also be written as follows: $-9 < x \leq 6$. Due to the graphs of their solution sets (shown below), compound inequalities such as these are referred to as conjunctions.

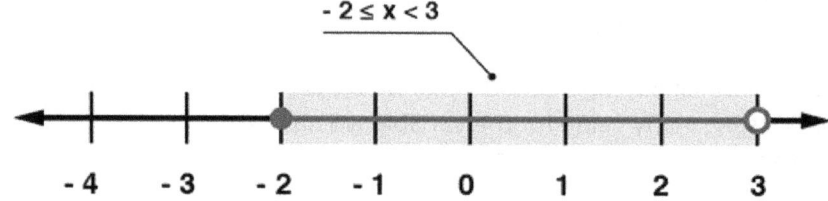

If there are no values that would make both inequalities of a compound inequality containing *and* true, then there is no solution. An example would be $x > 2$ and $x \leq 0$.

The solution set for a compound inequality containing *or* consists of all the values for the variable that make at least one of the inequalities true. The solution set for the compound inequality $x \leq -2$ or $x > 1$ consists of -2, all values less than -2, and all values greater than 1. Due to the graphs of their solution sets (shown below), compound inequalities such as these are referred to as disjunctions.

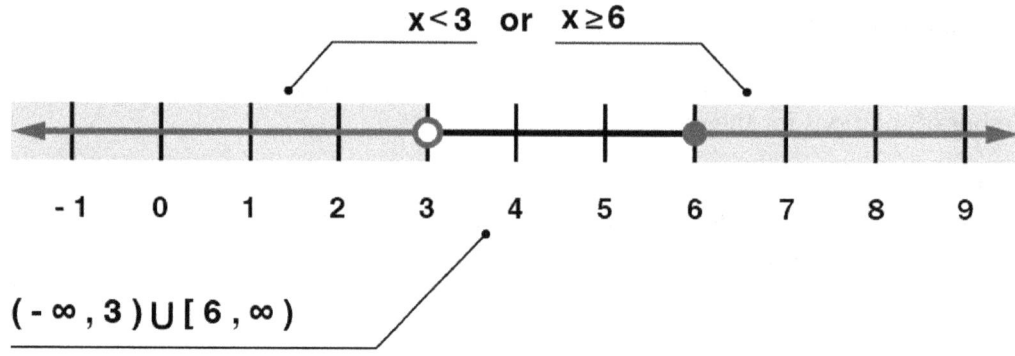

If the two inequalities for a compound inequality containing *or* "overlap," then the solution set contains all real numbers. An example would be $x > 2$ or $x < 7$. Any number would make at least one of these true.

Algebraically Solving Systems of Two Linear Equations in Two Variables

A system of two linear equations in two variables is a set of equations that use the same variables (typically *x* and *y*). A solution to the system is an ordered pair that makes both equations true. One method for solving a system is by graphing. This method, however, is not always practical. Students may not have graph paper; or the solution may not consist of integers, making it difficult to identify the exact point of intersection on a graph. There are two methods for solving systems of equations algebraically: substitution and elimination. The method used will depend on the characteristics of the equations in the system.

Solving Systems of Equations with the Substitution Method
If one of the equations in a system has an isolated variable (*x*= or *y*=) or a variable that can be easily isolated, the substitution method can be used. Here's a sample system: $x + 3y = 7; 2x - 4y = 24$. The first equation can easily be solved for *x*. By subtracting $3y$ on both sides, the resulting equation is $x = 7 - 3y$. When one equation is solved for a variable, the expression that it is equal can be substituted

into the other equation. For this example, $(7 - 3y)$ would be substituted for *x* into the second equation as follows:

$$2(7 - 3y) + 4y = 24$$

Solving this equation results in $y = -5$. Once the value for one variable is known, this value should be substituted into either of the original equations to determine the value of the other variable. For the example, -5 would be substituted for *y* in either of the original equations. Substituting into the first equation results in $x + 3(-5) = 7$, and solving this equation yields $x = 22$. The solution to a system is an ordered pair, so the solution to the example is written as (22, 7). The solution can be checked by substituting it into both equations of the system to ensure it results in two true statements.

<u>Solving Systems of Equations with the Elimination Method</u>
The elimination method for solving a system of equations involves canceling out (or eliminating) one of the variables. This method is typically used when both equations of a system are written in standard form $(Ax + By = C)$. An example is

$$2x + 3y = 12$$

$$5x - y = 13$$

To perform the elimination method, the equations in the system should be arranged vertically to be added together and then one or both of the equations should be multiplied so that one variable will be eliminated when the two are added. Opposites will cancel each other when added together. For example, 8x and -8x will cancel each other when added. For the example above, writing the system vertically helps identify that the bottom equation should be multiplied by 3 to eliminate the variable *y*.

$$2x + 3y = 12 \quad \rightarrow \quad 2x + 3y = 12$$

$$3(5x - y = 13) \quad \rightarrow \quad 15x - 3y = 39$$

Adding the two equations together vertically results in $17x = 51$. Solving yields $x = 3$. Once the value for one variable is known, it can be substituted into either of the original equations to determine the value of the other variable. Once this is obtained, the solution can be written as an ordered pair (*x, y*) and checked in both equations of the system. In this example, the solution is (3, 2).

<u>Systems of Equations with No Solution or an Infinite Number of Solutions</u>
A system of equations can have one solution, no solution, or an infinite number of solutions. If, while solving a system algebraically, both variables cancel out, then the system has either no solution or has an infinite number of solutions. If the remaining constants result in a true statement (ex., $7 = 7$), then there is an infinite number of solutions. This would indicate coinciding lines. If the remaining constants result in a false statement, then there is no solution to the system. This would indicate parallel lines.

Interpreting Variables and Constants in Expressions for Linear Functions in the Context Presented

Linear functions, also written as linear equations in two variables, can be written to model real-world scenarios. Questions on this material will provide information about a scenario and then request a linear equation to represent the scenario. The algebraic process for writing the equation will depend on the given information. The key to writing linear models is to decipher the information given to determine what it represents in the context of a linear equation (variables, slope, ordered pairs, etc.).

Identifying Variables for Linear Models

The first step to writing a linear model is to identify what the variables represent. A variable represents an unknown quantity, and in the case of a linear equation, a specific relationship exists between the two variables (usually x and y). Within a given scenario, the variables are the two quantities that are changing. The variable x is considered the independent variable and represents the inputs of a function. The variable y is considered the dependent variable and represents the outputs of a function. For example, if a scenario describes distance traveled and time traveled, distance would be represented by y and time represented by x. The distance traveled depends on the time spent traveling (time is independent). If a scenario describes the cost of a cab ride and the distance traveled, the cost would be represented by y and the distance represented by x. The cost of a cab ride depends on the distance traveled.

Identifying the Slope and Y-Intercept for Linear Models

The slope of the graph of a line represents the rate of change between the variables of an equation. In the context of a real-world scenario, the slope will tell the way in which the unknown quantities (variables) change with respect to each other. A scenario involving distance and time might state that someone is traveling at a rate of 45 miles per hour. The slope of the linear model would be 45. A scenario involving the cost of a cab ride and distance traveled might state that the person is charged $3 for each mile. The slope of the linear model would be 3.

The y-intercept of a linear function is the value of y when $x = 0$ (the point where the line intercepts the y-axis on the graph of the equation). It is sometimes helpful to think of this as a "starting point" for a linear function. Suppose for the scenario about the cab ride that the person is told that the cab company charges a flat fee of $5 plus $3 for each mile. Before traveling any distance ($x = 0$), the cost is $5. The y-intercept for the linear model would be 5.

Identifying Ordered Pairs for Linear Models

A linear equation with two variables can be written given a point (ordered pair) and the slope or given two points on a line. An ordered pair gives a set of corresponding values for the two variables (x and y). As an example, for a scenario involving distance and time, it is given that the person traveled 112.5 miles in 2 ½ hours. Knowing that x represents time and y represents distance, this information can be written as the ordered pair (2.5, 112.5).

Understanding Connections Between Algebraic and Graphical Representations

The solution set to a linear equation in two variables can be represented visually by a line graphed on the coordinate plane. Every point on this line represents an ordered pair (x, y), which makes the equation true. The process for graphing a line depends on the form in which its equation is written: slope-intercept form or standard form.

Graphing a Line in Slope-Intercept Form

When an equation is written in slope-intercept form, $y = mx + b$, m represents the slope of the line and b represents the y-intercept. The y-intercept is the value of y when $x = 0$ and the point at which the graph of the line crosses the y-axis. The slope is the rate of change between the variables, expressed as a fraction. The fraction expresses the change in y compared to the change in x. If the slope is an integer, it should be written as a fraction with a denominator of 1. For example, 5 would be written as 5/1.

To graph a line given an equation in slope-intercept form, the y-intercept should first be plotted. For example, to graph $y = -\frac{2}{3}x + 7$, the y-intercept of 7 would be plotted on the y-axis (vertical axis) at the point (0, 7). Next, the slope would be used to determine a second point for the line. Note that all that is necessary to graph a line is two points on that line. The slope will indicate how to get from one point on the line to another. The slope expresses vertical change (y) compared to horizontal change (x) and therefore is sometimes referred to as $\frac{rise}{run}$. The numerator indicates the change in the y value (move up for positive integers and move down for negative integers), and the denominator indicates the change in the x value. For the previous example, using the slope of $-\frac{2}{3}$, from the first point at the y-intercept, the second point should be found by counting down 2 and to the right 3. This point would be located at (3, 5).

Graphing a Line in Standard Form
When an equation is written in standard form, $Ax + By = C$, it is easy to identify the x- and y-intercepts for the graph of the line. Just as the y-intercept is the point at which the line intercepts the y-axis, the x-intercept is the point at which the line intercepts the x-axis. At the y-intercept, $x = 0$; and at the x-intercept, $y = 0$. Given an equation in standard form, $x = 0$ should be used to find the y-intercept. Likewise, $y = 0$ should be used to find the x-intercept. For example, to graph $3x + 2y = 6$, 0 for y results in $3x + 2(0) = 6$. Solving for y yields $x = 2$; therefore, an ordered pair for the line is (2, 0). Substituting 0 for x results in $3(0) + 2y = 6$. Solving for y yields $y = 3$; therefore, an ordered pair for the line is (0, 3). The two ordered pairs (the x- and y-intercepts) can be plotted, and a straight line through them can be constructed.

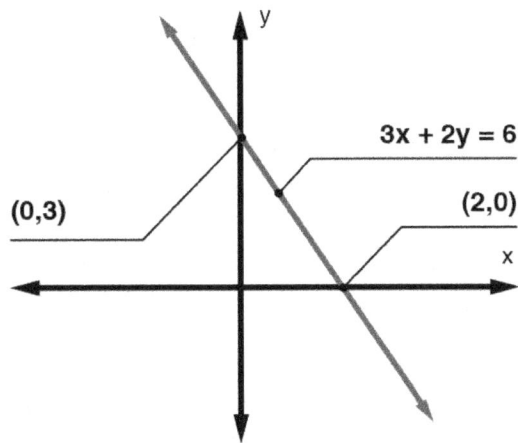

Writing the Equation of a Line Given its Graph

Given the graph of a line, its equation can be written in two ways. If the y-intercept is easily identified (is an integer), it and another point can be used to determine the slope. When determining $\frac{change\ in\ y}{change\ in\ x}$ from one point to another on the graph, the distance for $\frac{rise}{run}$ is being figured. The equation should be written in slope-intercept form, $y = mx + b$, with m representing the slope and b representing the y-intercept.

The equation of a line can also be written by identifying two points on the graph of the line. To do so, the slope is calculated and then the values are substituted for the slope and either of the ordered pairs into the point-slope form of an equation.

Vertical, Horizontal, Parallel, and Perpendicular Lines

For a vertical line, the value of x remains constant (for all ordered pairs (x, y) on the line, the value of x is the same); therefore, the equations for all vertical lines are written in the form $x = number$. For example, a vertical line that crosses the x-axis at -2 would have an equation of $x = -2$. For a horizontal line, the value of y remains constant; therefore, the equations for all horizontal lines are written in the form $y = number$.

Parallel lines extend in the same exact direction without ever meeting. Their equations have the same slopes and different y-intercepts. For example, given a line with an equation of $y = -3x + 2$, a parallel line would have a slope of -3 and a y-intercept of any value other than 2. Perpendicular lines intersect to form a right angle. Their equations have slopes that are opposite reciprocal (the sign is changed and the fraction is flipped; for example, $-\frac{2}{3}$ and $\frac{3}{2}$) and y-intercepts that may or may not be the same. For example, given a line with an equation of $y = \frac{1}{2}x + 7$, a perpendicular line would have a slope of $-\frac{2}{1}$ and any value for its y-intercept.

Problem-Solving and Data Analysis

Using Ratios, Rates, Proportions, and Scale Drawings to Solve Single- and Multistep Problems

Ratios, rates, proportions, and scale drawings are used when comparing two quantities. Questions on this material will include expressing relationships in simplest terms and solving for missing quantities.

Ratios

A ratio is a comparison of two quantities that represent separate groups. For example, if a recipe calls for 2 eggs for every 3 cups of milk, it can be expressed as a ratio. Ratios can be written three ways: (1) with the word "to"; (2) using a colon; or (3) as a fraction. For the previous example, the ratio of eggs to cups of milk can be written as: 2 to 3, 2:3, or $\frac{2}{3}$. When writing ratios, the order is important. The ratio of eggs to cups of milk is not the same as the ratio of cups of milk to eggs, 3:2.

In simplest form, both quantities of a ratio should be written as integers. These should also be reduced just as a fraction would be. For example, 5:10 would reduce to 1:2. Given a ratio where one or both quantities are expressed as a decimal or fraction, both should be multiplied by the same number to produce integers. To write the ratio $\frac{1}{3}$ to 2 in simplest form, both quantities should be multiplied by 3. The resulting ratio is 1 to 6.

When a problem involving ratios gives a comparison between two groups, then: (1) a total should be provided and a part should be requested; or (2) a part should be provided and a total should be requested. Consider the following:

> The ratio of boys to girls in the 11th grade is 5:4. If there is a total of 270 11th grade students, how many are girls?

To solve this, the total number of "ratio pieces" first needs to be determined. The total number of 11th grade students is divided into 9 pieces. The ratio of boys to total students is 5:9; and the ratio of girls to total students is 4:9. Knowing the total number of students, the number of girls can be determined by setting up a proportion: $\frac{4}{9} = \frac{x}{270}$. Solving the proportion, it shows that there are 120 11th grade girls.

Rates
A rate is a ratio comparing two quantities expressed in different units. A unit rate is one in which the second is one unit. Rates often include the word *per*. Examples include miles per hour, beats per minute, and price per pound. The word *per* can be represented with a / symbol or abbreviated with the letter "p" and the units abbreviated. For example, miles per hour would be written mi/h. Given a rate that is not in simplest form (second quantity is not one unit), both quantities should be divided by the value of the second quantity. Suppose a patient had 99 heartbeats in 1½ minutes. To determine the heart rate, 1½ should divide both quantities. The result is 66 bpm.

Scale Drawings
Scale drawings are used in designs to model the actual measurements of a real-world object. For example, the blueprint of a house might indicate that it is drawn at a scale of 3 inches to 8 feet. Given one value and asked to determine the width of the house, a proportion should be set up to solve the problem. Given the scale of 3in:8ft and a blueprint width of 1 ft (12 in.), to find the actual width of the building, the proportion $\frac{3}{8} = \frac{12}{x}$ should be used. This results in an actual width of 32 ft.

Proportions
A proportion is a statement consisting of two equal ratios. Proportions will typically give three of four quantities and require solving for the missing value. The key to solving proportions is to set them up properly. Here's a sample problem:

> If 7 gallons of gas costs $14.70, how many gallons can you get for $20?

The information should be written as equal ratios with a variable representing the missing quantity

$$\left(\frac{gallons}{cost} = \frac{gallons}{cost}\right) : \frac{7}{14.70} = \frac{x}{20}$$

To solve, cross multiply (multiply the numerator of the first ratio by the denominator of the second and vice versa) is used and the products are set equal to each other. Cross-multiplying results in: $(7)(20) = (14.7)(x)$. Solving the equation for x, it can be determined that 9.5 gallons of gas can be purchased for $20.

Indirect Proportions
The proportions described above are referred to as direct proportions or direct variation. For direct proportions, as one quantity increases, the other quantity also increases. For indirect proportions (also referred to as indirect variations, inverse proportions, or inverse variations), as one quantity increases,

the other decreases. Direct proportions can be written: $\frac{y_1}{x_1} = \frac{y_2}{x_2}$. Conversely, indirect proportions are written: $y_1 x_1 = y_2 x_2$. Here's a sample problem:

> It takes 3 carpenters 10 days to build the frame of a house. How long should it take 5 carpenters to build the same frame?

In this scenario, as one quantity increases (number of carpenters), the other decreases (number of days building); therefore, this is an inverse proportion. To solve, the products of the two variables (in this scenario, the total work performed) are set equal to each other ($y_1 x_1 = y_2 x_2$). Using y to represent carpenters and x to represent days, the resulting equation is: $(3)(10) = (5)(x2)$. Solving for x_2, it is determined that it should take 5 carpenters 6 days to build the frame of the house.

Solving Single- and Multistep Problems Involving Percentages

The word percent means "per hundred." When dealing with percentages, it may be helpful to think of the number as a value in hundredths. For example, 15% can be expressed as "fifteen hundredths" and written as $\frac{15}{100}$ or .15.

Converting from Decimals and Fractions to Percentages

To convert a decimal to a percent, a number is multiplied by 100. To write .25 as a percent, the equation $.25 x 100$ yields 25%. To convert a fraction to a percent, the fraction is converted to a decimal and then multiplied by 100. To convert $\frac{3}{5}$ to a decimal, the numerator (3) is divided by the denominator (5). This results in .6, which is then multiplied by 100 to get 60%.

To convert a percent to a decimal, the number is divided by 100. For example, 150% is equal to 1.5 $\left(\frac{150}{100}\right)$. To convert a percent to a fraction, the percent sign is deleted and the value is written as the numerator with a denominator of 100. For example, 2% = $\frac{2}{100}$. Fractions should be reduced: $\frac{2}{100} = \frac{1}{50}$.

Percent Problems

Material on percentages can include questions such as: What is 15% of 25? What percent of 45 is 3? Five is $\frac{1}{2}$% of what number? To solve these problems, the information should be rewritten as an equation where the following helpful steps are completed: (1) "what" is represented by a variable (x); (2) "is" is represented by an = sign; and (3) "of" is represented by multiplication. Any values expressed as a percent should be written as a decimal; and if the question is asking for a percent, the answer should be converted accordingly. Here are three sample problems based on the information above:

What is 15% of 25?	What percent of 45 is 3?	Five is $\frac{1}{2}$% of what number?
$x = .15 \times 25$	$x \times 45 = 3$	$5 = .005 \times x$
$x = 3.75$	$x = 0.0\bar{6}$	$x = 1,000$
	$x = 6.\bar{6}\%$	

Percent Increase/Decrease

Problems dealing with percentages may involve an original value, a change in that value, and a percentage change. A problem will provide two pieces of information and ask to find the third. To do so, this formula is used: $\frac{change}{original\ value} \times 100 =$ percent change. Here's a sample problem:

> Attendance at a baseball stadium has dropped 16% from last year. Last year's average attendance was 40,000. What is this year's average attendance?

Using the formula and information, the change is unknown (x), the original value is 40,000, and the percent change is 16%. The formula can be written as: $\frac{x}{40,000} \times 100 = 16$. When solving for x, it is determined the change was 6,400. The problem asked for this year's average attendance, so to calculate, the change (6,400) is subtracted from last year's attendance (40,000) to determine this year's average attendance is 33,600.

Percent More Than/Less Than

Percentage problems may give a value and what percent that given value is more than or less than an original unknown value. Here's a sample problem:

> A store advertises that all its merchandise has been reduced by 25%. The new price of a pair of shoes is $60. What was the original price?

This problem can be solved by writing a proportion. Two ratios should be written comparing the cost and the percent of the original cost. The new cost is 75% of the original cost (100% - 25%); and the original cost is 100% of the original cost. The unknown original cost can be represented by x. The proportion would be set up as: $\frac{60}{75} = \frac{x}{100}$. Solving the proportion, it is determined the original cost was $80.

Solving Single- and Multistep Problems Involving Measurement Quantities, Units, and Unit Conversion

Unit Rates

A rate is a ratio in which two terms are in different units. When rates are expressed as a quantity of one, they are considered unit rates. To determine a unit rate, the first quantity is divided by the second. Knowing a unit rate makes calculations easier than simply having a rate. Suppose someone bought a 3lb bag of onions for $1.77. To calculate the price of 5lbs of onions, a proportion could be set up as follows:

$$\frac{3}{1.77} = \frac{5}{x}$$

However, knowing the unit rate, multiplying the value of pounds of onions by the unit price is another way to find the solution: (The unit price would be calculated $1.77/3lb = $0.59/lb.)

$$5lbs \times \frac{\$.59}{lb} = \$2.95. \text{ (The "lbs" units cancel out.)}$$

Unit Conversion

Unit conversions apply to many real-world scenarios, including cooking, measurement, construction, and currency. Problems on this material can be solved similarly to those involving unit rates. Given the conversion rate, it can be written as a fraction (ratio) and multiplied by a quantity in one unit to convert it to the corresponding unit. For example, someone might want to know how many minutes are in 3½ hours. The conversion rate of 60 minutes to 1 hour can be written as $\frac{60\ min}{1\ h}$. Multiplying the quantity by the conversion rate results in $3\frac{1}{2}h \times \frac{60\ min}{1\ h} = 210\ min$. The "h" unit is canceled. To convert a quantity in minutes to hours, the fraction for the conversion rate would be flipped (to cancel the "min" unit). To convert 195 minutes to hours, the equation $195\ min \times \frac{1h}{60min}$ would be used. The result is $\frac{195h}{60}$, which reduces to $3\frac{1}{4}$ hours.

Converting units may require more than one multiplication. The key is to set up the conversion rates so that units cancel out each other and the desired unit is left. Suppose someone wants to convert 3.25 yards to inches, given that 1yd = 3ft and 12in = 1ft. To calculate, the equation $3.25yd \times \frac{3ft}{1yd} \times \frac{12in}{1ft}$ would be used. The "yd" and "ft" units will cancel, resulting in 117 inches.

Given a Scatterplot, Using Linear, Quadratic, or Exponential Models to Describe How Variables are Related

Scatterplots can be used to determine whether a correlation exists between two variables. The horizontal (x) axis represents the independent variable and the vertical (y) axis represents the dependent variable. If when graphed, the points model a linear, quadratic, or exponential relationship, then a correlation is said to exist. If so, a line of best-fit or curve of best-fit can be drawn through the points, with the points relatively close on either side. Writing the equation for the line or curve allows for predicting values for the variables. Suppose a scatterplot displays the value of an investment as a function of years after investing. By writing an equation for the line or curve and substituting a value for one variable into the equation, the corresponding value for the other variable can be calculated.

Linear Models

If the points of a scatterplot model a linear relationship, a line of best-fit is drawn through the points. If the line of best-fit has a positive slope (y-values increase as x-values increase), then the variables have a positive correlation. If the line of best-fit has a negative slope (y-values decrease as x-values increase), then a negative correlation exists. A positive or negative correlation can also be categorized as strong or weak, depending on how closely the points are grouped around the line of best-fit.

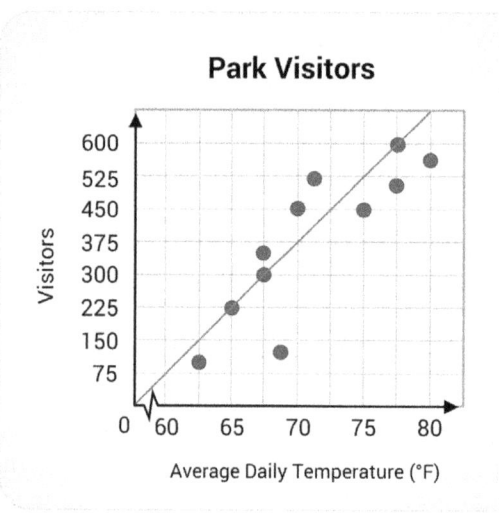

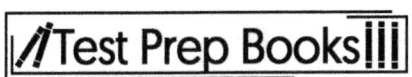

Given a line of best-fit, its equation can be written by identifying: the slope and *y*-intercept; a point and the slope; or two points on the line.

Quadratic Models

A quadratic function can be written in the form $y = ax^2 + bx + c$. The u-shaped graph of a quadratic function is called a parabola. The graph can either open up or open down (upside down u). The graph is symmetric about a vertical line, called the axis of symmetry. Corresponding points on the parabola are directly across from each other (same *y*-value) and are the same distance from the axis of symmetry (on either side). The axis of symmetry intersects the parabola at its vertex. The *y*-value of the vertex represents the minimum or maximum value of the function. If the graph opens up, the value of *a* in its equation is positive, and the vertex represents the minimum of the function. If the graph opens down, the value of *a* in its equation is negative, and the vertex represents the maximum of the function.

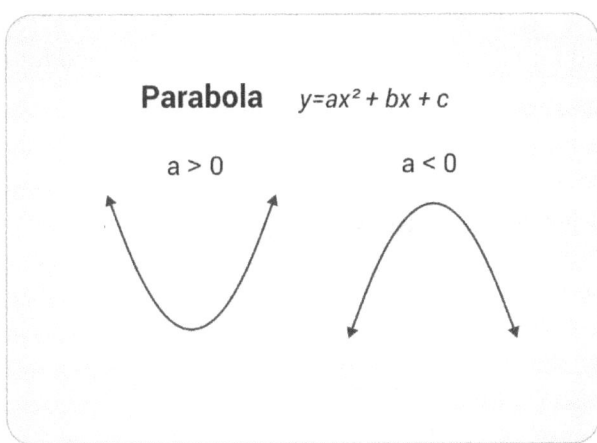

Given a curve of best-fit that models a quadratic relationship, the equation of the parabola can be written by identifying the vertex of the parabola and another point on the graph. The values for the vertex (h, k) and the point (x, y) should be substituted into the vertex form of a quadratic function, $y = a(x - h)^2 - k$, to determine the value of *a*. To write the equation of a quadratic function with a vertex of (4, 7) and containing the point (8, 3), the values for *h*, *k*, *x*, and *y* should be substituted into the vertex form of a quadratic function, resulting in

$$3 = a(8 - 4)^2 + 7$$

Solving for *a*, yields $a = -\frac{1}{4}$. Therefore, the equation of the function can be written as

$$y = -\frac{1}{4}(x - 4)^2 + 7$$

The vertex form can be manipulated in order to write the quadratic function in standard form.

Exponential Models

An exponential curve can be used as a curve of best-fit for a scatterplot. The general form for an exponential function is $y = ab^x$ where *b* must be a positive number and cannot equal 1. When the value of *b* is greater than 1, the function models exponential growth (as *x* increases, *y* increases). When the value of *b* is less than 1, the function models exponential decay (as *x* increases, *y* decreases). If *a* is positive, the graph consists of points above the x-axis; and if *a* is negative, the graph consists of points below the x-axis. An asymptote is a line that a graph approaches.

Given a curve of best-fit modeling an exponential function, its equation can be written by identifying two points on the curve. To write the equation of an exponential function containing the ordered pairs (2, 2) and (3, 4), the ordered pair (2, 2) should be substituted in the general form and solved for *a*:

$$2 = a \times b^2 \rightarrow a = \frac{2}{b^2}$$

The ordered pair (3, 4) and $\frac{2}{b^2}$ should be substituted in the general form and solved for *b*:

$$4 = \frac{2}{b^2} \times b^3 \rightarrow b = 2$$

Then, 2 should be substituted for *b* in the equation for *a* and then solved for *a*: $a = \frac{2}{2^2} \rightarrow a = \frac{1}{2}$. Knowing the values of *a* and *b*, the equation can be written as: $y = \frac{1}{2} \times 2^x$.

Exponential Curve

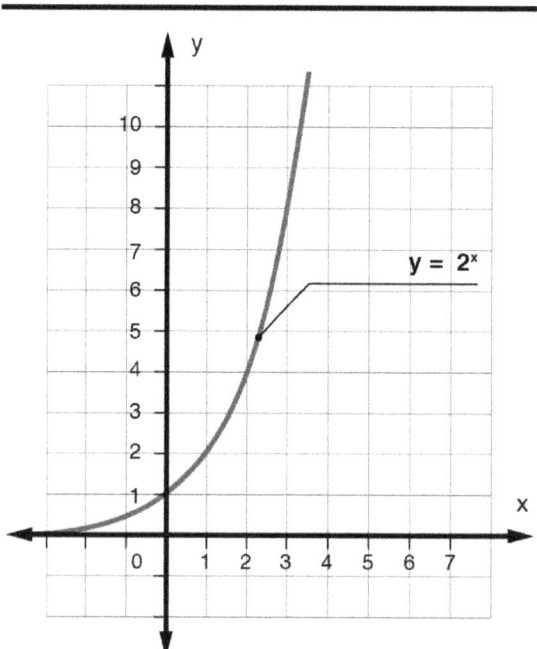

Using the Relationship Between Two Variables to Investigate Key Features of a Graph

Material on graphing relationships between two variables may include linear, quadratic, and exponential functions.

Graphing Quadratic Functions

The standard form of a quadratic function is $y = ax^2 + bx + c$. The graph of a quadratic function is a u-shaped (or upside down u) curve, called a parabola, which is symmetric about a vertical line (axis of symmetry). To graph a parabola, its vertex (high or low point for the curve) and at least two points on each side of the axis of symmetry need to be determined.

Given a quadratic function in standard form, the axis of symmetry for its graph is the line $x = -\frac{b}{2a}$. The vertex for the parabola has an x-coordinate of $-\frac{b}{2a}$. To find the y-coordinate for the vertex, the calculated x-coordinate needs to be substituted. To complete the graph, two different x-values need to be selected and substituted into the quadratic function to obtain the corresponding y-values. This will give two points on the parabola. These two points and the axis of symmetry are used to determine the two points corresponding to these. The corresponding points are the same distance from the axis of symmetry (on the other side) and contain the same y-coordinate. Plotting the vertex and four other points on the parabola allows for constructing the curve.

Quadratic Function

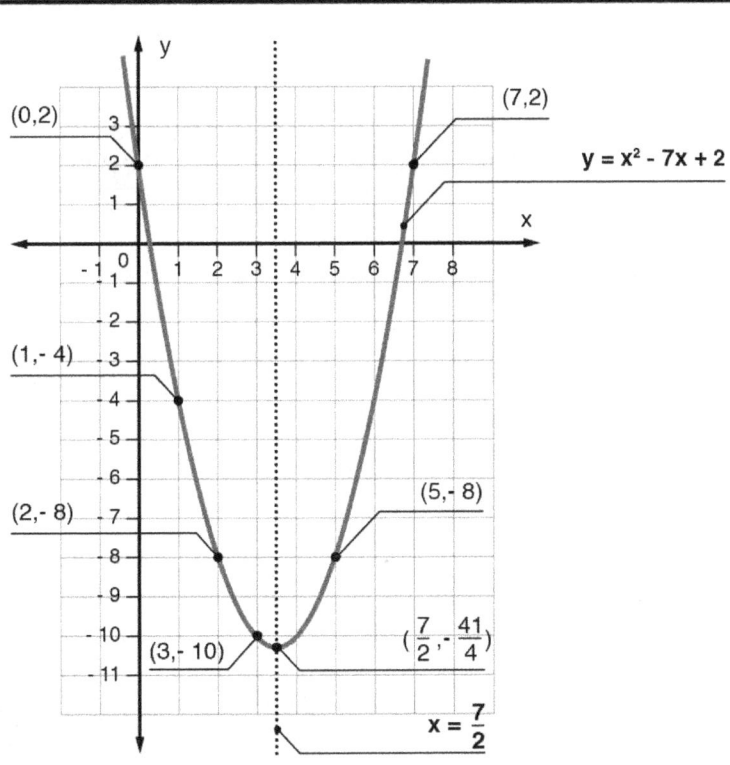

Graphing Exponential Functions

Exponential functions have a general form of $y = a \times b^x$. The graph of an exponential function is a curve that slopes upward or downward from left to right. The graph approaches a line, called an asymptote, as *x* or *y* increases or decreases. To graph the curve for an exponential function, *x*-values are selected and then substituted into the function to obtain the corresponding *y*-values. A general rule of thumb is to select three negative values, zero, and three positive values. Plotting the seven points on the graph for an exponential function should allow for constructing a smooth curve through them.

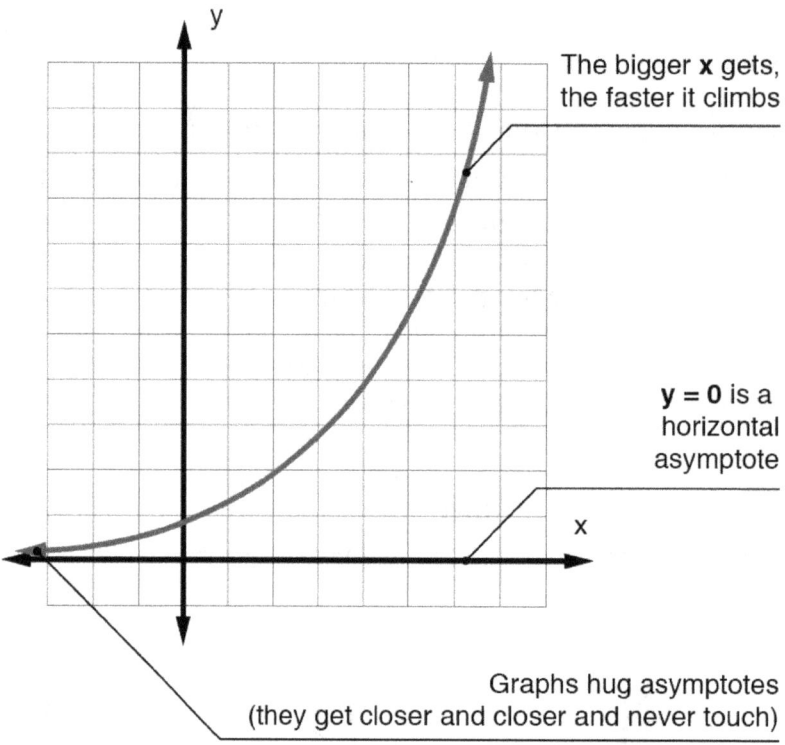

Comparing Linear Growth with Exponential Growth

Both linear and exponential equations can model a relationship of growth or decay between two variables. If the dependent variable (*y*) increases as the independent variable (*x*) increases, the relationship is referred to as growth. If *y* decreases as *x* increases, the relationship is referred to as decay.

Linear Growth and Decay

A linear function can be written in the form $y = mx + b$, where *x* represents the inputs, *y* represents the outputs, *b* represents the *y*-intercept for the graph, and *m* represents the slope of the line. The *y*-intercept is the value of *y* when $x = 0$ and can be thought of as the "starting point." The slope is the rate of change between the variables *x* and *y*. A positive slope represents growth; and a negative slope represents decay. Given a table of values for inputs (*x*) and outputs (*y*), a linear function would model the relationship if: *x* and *y* change at a constant rate per unit interval—for every two inputs a given distance apart, the distance between their corresponding outputs is constant. Here are some sample ordered pairs:

x	0	1	2	3
y	-7	-4	-1	2

For every 1 unit increase in x, y increases by 3 units. Therefore, the change is constant and thus represents linear growth.

Given a scenario involving growth or decay, determining if there is a constant rate of change between inputs (x) and outputs (y) will identify if a linear model is appropriate. A scenario involving distance and time might state that someone is traveling at a rate of 45 miles per hour. For every hour traveled (input), the distance traveled (output) increases by 45 miles. This is a constant rate of change.

Exponential Growth and Decay

An exponential function can be written in the form $y = a \times b^x$, where x represents the inputs, y represents the outputs, a represents the y-intercept for the graph, and b represents the growth rate. The y-intercept is the value of y when $x = 0$ and can be thought of as the "starting point." If b is greater than 1, the function describes exponential growth; and if b is less than 1, the function describes exponential decay. Given a table of values for inputs (x) and outputs (y), a linear function would model the relationship if the variables change by a common ratio over given intervals—for every two inputs a given distance apart, the quotients of their corresponding outputs is constant. Here are some sample ordered pairs:

x	0	1	2	3
y	3	6	12	24

For every 1 unit increase in x, the quotient of the corresponding y-values equals $\frac{1}{2}$ (ex., $\frac{3}{6}, \frac{6}{12}, \frac{12}{24}$). Therefore, the table represents exponential growth.

Given a scenario describing an exponential function, the growth or decay is expressed using multiplication. Words such as "doubling" and "halving" will often be used. A problem might indicate that the value of an investment triples every year or that every decade the population of an insect is halved. These indicate exponential growth and decay.

Using Two-Way Tables to Summarize Categorical Data and Relative Frequencies, and Calculate Conditional Probability

Categorical data consists of numerical values found by dividing the entire set into subsets based on variables that represent categories. An example would be the survey results of high school seniors, specifying gender and asking whether they consume alcohol. The data can be arranged in a two-way frequency table (also called a contingency table).

Two-Way Frequency/Contingency Tables

A contingency table presents the frequency tables of both variables simultaneously, as shown below. The levels of one variable constitute the rows of the table, and the levels of the other constitute the columns. The margins consist of the sum of cell frequencies for each row and each column (marginal frequencies). The lower right corner value is the sum of marginal frequencies for the rows or the sum of the marginal frequencies for the columns. Both are equal to the total sample size.

	Drink Alcohol	Do Not Drink Alcohol	Total
Male	63	51	114
Female	37	68	105
Total	100	119	219

Conditional Frequencies

To calculate a conditional relative frequency, the cell frequency is divided by the marginal frequency for the desired outcome given the conditional category. For instance, using the table to determine the relative frequency that a female drinks, the number of females who drink (desired outcome) is divided by the total number of females (conditional category). The conditional relative frequency would equal $\frac{37}{105}$, which equals .35. If a problem asks for a conditional probability, the answer would be expressed as a fraction in simplest form. If asked for a percent, multiply the decimal by 100.

Association of Variables

An association between the variables exists if the conditional relative frequencies are different depending on condition. If the conditional relative frequencies are close to equal, then the variables are independent. For our example, 55% of senior males and 35% of senior females drink alcohol. The difference between frequencies across conditions (male or female) is enough to conclude that an association exists between the variables.

Making Inferences about Population Parameters Based on Sample Data

Statistical inference, based in probability theory, makes calculated assumptions about an entire population based on data from a sample set from that population.

Population Parameters

A population is the entire set of people or things of interest. Suppose a study is intended to determine the number of hours of sleep per night for college females in the U.S. The population would consist of EVERY college female in the country. A sample is a subset of the population that may be used for the study. It would not be practical to survey every female college student, so a sample might consist of 100 students per school from 20 different colleges in the country. From the results of the survey, a sample statistic can be calculated. A sample statistic is a numerical characteristic of the sample data, including mean and variance. A sample statistic can be used to estimate a corresponding population parameter. A population parameter is a numerical characteristic of the entire population. Suppose the sample data had a mean (average) of 5.5. This sample statistic can be used as an estimate of the population parameter (average hours of sleep for every college female in the U.S.).

Confidence Intervals

A population parameter is usually unknown and therefore is estimated using a sample statistic. This estimate may be highly accurate or relatively inaccurate based on errors in sampling. A confidence interval indicates a range of values likely to include the true population parameter. These are constructed at a given confidence level, such as 95%. This means that if the same population is sampled repeatedly, the true population parameter would occur within the interval for 95% of the samples.

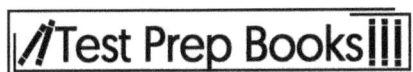

Measurement Error

The accuracy of a population parameter based on a sample statistic may also be affected by measurement error, which is the difference between a quantity's true value and its measured value. Measurement error can be divided into random error and systematic error. An example of random error for the previous scenario would be a student reporting 8 hours of sleep when she actually sleeps 7 hours per night. Systematic errors are those attributed to the measurement system. Suppose the sleep survey gave response options of 2, 4, 6, 8, or 10 hours. This would lead to systematic measurement error.

Using Statistics to Investigate Measures of Center of Data and Analyzing Shape, Center, and Spread

Descriptive statistics are used to gain an understanding of properties of a data set. This entails examining the center, spread, and shape of the sample data.

Center

The center of the sample set can be represented by its mean, median, or mode. The mean is the average of the data set, calculated by adding the data values and dividing by the sample size. The median is the value of the data point in the middle when the sample is arranged in numerical order. If the sample has an even number of data points, the mean of the two middle values is the median. The mode is the value that appears most often in a data set. It is possible to have multiple modes (if different values repeat equally as often) or no mode (if no value repeats).

Spread

Methods for determining the spread of the sample include calculating the range and standard deviation for the data. The range is calculated by subtracting the lowest value from the highest value in the set. The standard deviation of the sample can be calculated using the formula:

$$\sigma = \sqrt{\frac{\sum(x - \bar{x})^2}{n - 1}}$$

where $\bar{x}$ = sample mean and n = sample size.

Shape

The shape of the sample when displayed as a histogram or frequency distribution plot helps to determine if the sample is normally distributed (bell-shaped curve), symmetrical, or has measures of skewness (lack of symmetry) or kurtosis. Kurtosis is a measure of whether the data are heavy-tailed (high number of outliers) or light-tailed (low number of outliers).

Evaluating Reports to Make Inferences, Justify Conclusions, and Determine Appropriateness of Data Collection Methods

The presentation of statistics can be manipulated to produce a desired outcome. Here's a statement to consider: "Four out of five dentists recommend our toothpaste." Who are the five dentists? This statement is very different from the statement: "Four out of every five dentists recommend our toothpaste." Whether intentional or unintentional, statistics can be misleading. Statistical reports should be examined to verify the validity and significance of the results. The context of the numerical values allows for deciphering the meaning, intent, and significance of the survey or study. Questions on this material will require students to use critical thinking skills to justify or reject results and conclusions.

When analyzing a report, who conducted the study and their intent should be considered. Was it performed by a neutral party or by a person or group with a vested interest? A study on health risks of smoking performed by a health insurance company would have a much different intent than one performed by a cigarette company. The sampling method and the data collection method should be considered too. Was it a true random sample of the population or was one subgroup over- or underrepresented? If all 20 schools included in the study were state colleges, the results may be biased due to a lack of private school participants. Also, the measurement system used to obtain the data should be noted. Was the system accurate and precise or was it a flawed system? If possible responses were limited for the sleep study to 2, 4, 6, 8, or 10, it could be argued that the measurement system was flawed.

Every scenario involving statistical reports will be different. The key is to examine all aspects of the study before determining whether to accept or reject the results and corresponding conclusions.

Passport to Advanced Math

Creating a Quadratic or Exponential Function

Quadratic Models

A quadratic function can be written in the standard form: $y = ax^2 + bx + c$. It can be represented by a u-shaped graph called a parabola. For a quadratic function where the value of a is positive, as the inputs increase, the outputs increase until a certain value (maximum of the function) is reached. As inputs increase past the value that corresponds with the maximum output, the relationship reverses and the outputs decrease. For a quadratic function where a is negative, as the inputs increase, the outputs (1) decrease, (2) reach a maximum, and (3) then increase.

Consider a ball thrown straight up into the air. As time passes, the height of the ball increases until it reaches its maximum height. After reaching the maximum height, as time increases, the height of the ball decreases (it is falling toward the ground). This relationship can be expressed as a quadratic function where time is the input (x), and the height of the ball is the output (y).

Given a scenario that can be modeled by a quadratic function, to write its equation, the following is needed: its vertex and any other ordered pair; or any three ordered pairs for the function. Given three ordered pairs, they should be substituted into the general form ($y = ax^2 + bx + c$) to create a system of three equations. For example, given the ordered pairs (2, 3), (3, 13), and (4, 29), it yields:

$$3 = a(2)2 + b(2) + c \rightarrow 4a + 2b + c = 3$$

$$13 = a(3)2 + b(3) + c \rightarrow 9a + 3b + c = 13$$

$$29 = a(4)2 + b(4) + c \rightarrow 16a + 24b + c = 29$$

The values for a, b, and c in the system can be found and substituted into the general form to write the equation of the function. In this case, the equation is $y = 3x^2 - 5x + 1$.

Exponential Models

Exponential functions can be written in the form: $y = a \times b^x$. Scenarios involving growth and decay can be modeled by exponential functions.

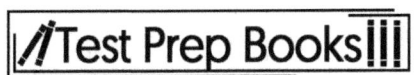

The equation for an exponential function can be written given the y-intercept (a) and the growth rate (b). The y-intercept is the output (y) when the input (x) equals zero. It can be thought of as an "original value" or starting point. The value of b is the rate at which the original value increases ($b > 1$) or decreases ($b < 1$). Suppose someone deposits $1200 into a bank account that accrues 1% interest per month. The y-intercept, a, would be $1200, while the growth rate, b, would be 1.01 (100% of the original value + 1% interest). This scenario could be written as the exponential function $y = 1200 \times 1.01^x$, where x represents the number of months since the deposit and y represents money in the account.

Given a scenario that models an exponential function, the equation can also be written when provided two ordered pairs.

Determining the Most Suitable Form of an Expression

It is possible for algebraic expressions and equations to be written that look completely different, yet are still equivalent. For instance, the expression $4(2x - 3) - 3x + 5$ is equivalent to the expression $5x - 7$. Given two algebraic expressions, it can be determined if they are equivalent by writing them in simplest form. Distribution should be used, if applicable, and like terms should be combined. Given two algebraic equations, it can be determined if they are equivalent by solving each for the same variable. Here are two sample equations to consider: $3x - 4y = 7$ and $x + 2 = \frac{4}{3}y + 4\frac{1}{3}$. To determine if they are equivalent, solving for x is required.

$$3x - 4y = 7 \qquad\qquad x + 2 = \frac{4}{3}y + 4\frac{1}{3}$$

$$3x = 4y + 7 \qquad\qquad x = \frac{4}{3}y + 2\frac{1}{3}$$

$$x = \frac{4}{3}y + \frac{7}{3} \qquad\qquad x = \frac{4}{3}y + 2\frac{1}{3}$$

The equations are equivalent.

Equivalent Forms of Functions

Equations in two variables can often be written in different forms to easily recognize a given trait of the function or its graph. Linear equations written in slope-intercept form allow for recognition of the slope and y-intercept; and linear equations written in standard form allow for identification of the x and y-intercepts. Quadratic functions written in standard form allow for identification of the y-intercept and for easy calculation of outputs; and quadratic functions written in vertex form allow for identification of the function's minimum or maximum output and its graph's vertex. Polynomial functions written in factored form allow for identification of the zeros of the function.

The method of substituting the same inputs (x-values) into functions to determine if they produce the same outputs can reveal if functions are not equivalent (different outputs). However, corresponding inputs and outputs do not necessarily indicate equivalent functions.

Creating Equivalent Expressions Involving Rational Exponents

Converting To and From Radical Form

Algebraic expressions involving radicals ($\sqrt{x}$, $\sqrt[3]{x}$, etc.) can be written without the radical by using rational (fraction) exponents. For radical expressions, the value under the root symbol is called the radicand, and the type of root determines the index. For example, the expression $\sqrt{6x}$ has a radicand of 6x and index of 2 (it is a square root). If the exponent of the radicand is 1, then $\sqrt[n]{a} = a^{\frac{1}{n}}$ where n is the index. A number or variable without a power has an implied exponent of 1. For example:

$$\sqrt{6} = 6^{\frac{1}{2}}$$

and

$$125^{\frac{1}{3}} = \sqrt[3]{125}$$

For any exponent of the radicand, $\sqrt[n]{a^m} = \left(\sqrt[n]{a}\right)^m = a^{\frac{m}{n}}$.

For example, $64^{\frac{5}{3}} = \sqrt[3]{64^5} \text{ or } \left(\sqrt[3]{64}\right)^5$; and $(xy)^{\frac{2}{3}} = \sqrt[3]{(xy)^2} \text{ or } \left(\sqrt[3]{xy}\right)^2$.

Simplifying Expressions with Rational Exponents

When simplifying expressions with rational exponents, all basic properties for exponents hold true. When multiplying powers of the same base (same value with or without the same exponent), the exponents are added.

For example:

$$x^{\frac{2}{7}} \times x^{\frac{3}{14}} = x^{\frac{1}{2}} \left(\frac{2}{7} + \frac{3}{14} = \frac{1}{2}\right)$$

When dividing powers of the same base, the exponents are subtracted. For example:

$$\frac{5^{\frac{2}{3}}}{5^{\frac{1}{2}}} = 5^{\frac{1}{6}} \left(\frac{2}{3} - \frac{1}{2} = \frac{1}{6}\right)$$

When raising a power to a power, the exponents are multiplied. For example:

$$\left(5^{\frac{1}{2}}\right)^4 = 5^2 \left(\frac{1}{2} \times 4 = 2\right)$$

When simplifying expressions with exponents, a number should never be raised to a power or a negative exponent. If a number has an integer exponent, its value should be determined. If the number has a rational exponent, it should be rewritten as a radical and the value determined if possible. A base with a negative exponent moves from the numerator to the denominator of a fraction (or vice versa) and is written with a positive exponent. For example, $x^{-3} = \frac{1}{x^3}$ and $\frac{2}{5x^{-2}} = \frac{2x^2}{5}$. The exponent of 5 is 1, and therefore the 5 does not move.

Here's a sample expression: $(27x^{-9})^{\frac{1}{3}}$. After the implied exponents are noted, a power should be raised to a power by multiplying exponents, which yields $27^{\frac{1}{3}}x^{-3}$. Next, the negative exponent is eliminated by moving the base and power: $\frac{27^{\frac{1}{3}}}{x^3}$. Then the value of the number is determined to a power by writing it in radical form: $\frac{\sqrt[3]{27}}{x^3}$. Simplifying yields $\frac{3}{x^3}$.

Creating an Equivalent Form of an Algebraic Expression

There are many different ways to write algebraic expressions and equations that are equivalent to each other. Converting expressions from standard form to factored form and vice versa are skills commonly used in advanced mathematics. Standard form of an expression arranges terms with variables powers in descending order (highest exponent to lowest and then constants). Factored form displays an expression as the product of its factors (what can be multiplied to produce the expression).

Converting Standard Form to Factored Form

To factor an expression, a greatest common factor needs to be factored out first. Then, if possible, the remaining expression needs to be factored into the product of binomials. A binomial is an expression with two terms.

Greatest Common Factor

The greatest common factor (GCF) of a monomial (one term) consists of the largest number that divides evenly into all coefficients (number part of a term); and if all terms contain the same variable, the variable with the lowest exponent. The GCF of

$$3x^4 - 9x^3 + 12x^2$$

would be $3x^2$. To write the factored expression, every term needs to be divided by the GCF, then the product of the resulting quotient and the GCF (using parentheses to show multiplication) should be written. For the previous example, the factored expression would be $3x^2(x^2 - 3x + 4)$.

Factoring $Ax^2 + Bx + C$ When $A = 1$

To factor a quadratic expression in standard form when the value of a is equal to 1, the factors that multiply to equal the value of c should be found and then added to equal the value of b (the signs of b and c should be included). The factored form for the expression will be the product of binomials:

$$(x + factor1)(x + factor2)$$

Here's a sample expression: $x^2 - 4x - 5$. The two factors that multiply to equal c(-5) and add together to equal b(-4) are -5 and 1. Therefore, the factored expression would be $(x - 5)(x + 1)$. Note $(x + 1)(x - 5)$ is equivalent.

Factoring a Difference of Squares

A difference of squares is a binomial expression where both terms are perfect squares (perfect square-perfect square). Perfect squares include 1, 4, 9, 16 . . . and x^2, x^4, x^6 . . .

The factored form of a difference of squares will be:

$$(\sqrt{term1} + \sqrt{term2})(\sqrt{term1} - \sqrt{term2})$$

For example:

$$x^2 - 4 = (x + 2)(x - 2)$$

And

$$25x^6 - 81 = (5x^3 + 9)(5x^3 - 9)$$

Factoring $Ax^2 + Bx + C$ when $A \neq 1$

To factor a quadratic expression in standard form when the value of a is not equal to 1, the factors that multiply to equal the value of $a \times c$ should be found and then added to equal the value of b. Next, the expression splitting the bx term should be rewritten using those factors. Instead of three terms, there will now be four. Then the first two terms should be factored using GCF, and a common binomial should be factored from the last two terms. The factored form will be: (common binomial) (2 terms out of binomials). In the sample expression:

$$2x^2 + 11x + 12$$

the value of $a \times c$ $(2x12) = 24$. Two factors that multiply to 24 and added together to yield b(11) are 8 and 3. The bx term (11x) can be rewritten by splitting it into the factors:

$$2x^2 + 8x + 3x + 12$$

A GCF from the first two terms can be factored as:

$$2x(x + 4) + 3x + 12$$

A common binomial from the last two terms can then be factored as:

$$2(x + 4) + 3(x + 4)$$

The factored form can be written as a product of binomials: $(x + 4)(2x + 3)$.

Converting Factored Form to Standard Form

To convert an expression from factored form to standard form, the factors are multiplied.

Solving a Quadratic Equation

A quadratic equation is one in which the highest exponent of the variable is 2. A quadratic equation can have two, one, or zero real solutions. Depending on its structure, a quadratic equation can be solved by (1) factoring, (2) taking square roots, or (3) using the quadratic formula.

Solving Quadratic Equations by Factoring

To solve a quadratic equation by factoring, the equation should first be manipulated to set the quadratic expression equal to zero. Next, the quadratic expression should be factored using the appropriate method(s). Then each factor should be set equal to zero. If two factors multiply to equal zero, then one or both factors must equal zero. Finally, each equation should be solved. Here's a sample:

$$x^2 - 10 = 3x - 6$$

The expression should be set equal to zero: $x^2 - 3x - 4 = 0$. The expression should be factored:

$$(x - 4)(x + 1) = 0$$

Each factor should be set equal to zero: $x - 4 = 0$; $x + 1 = 0$. Solving yields $x = 4$ or $x = -1$.

Solving Quadratic Equations by Taking Square Roots

If a quadratic equation does not have a linear term (variable to the first power), it can be solved by taking square roots. This means x^2 needs to be isolated and then the square root of both sides of the equation should be isolated. There will be two solutions because square roots can be positive or negative. ($\sqrt{4}$ = 2 or -2 because $2 \times 2 = 4$ and $-2 \times -2 = 4$.) Here's a sample equation: $3x^2 - 12 = 0$. Isolating x^2 yields $x^2 = 4$. The square root of both sides is then solved: $x = 2$ or -2.

The Quadratic Formula

When a quadratic expression cannot be factored or is difficult to factor, the quadratic formula can be used to solve the equation. To do so, the equation must be in the form:

$$ax^2 + bx + c = 0$$

The quadratic formula is:

$$x = \frac{-b \pm \sqrt{b^2 - 4ac}}{2a}$$

(The $\pm$ symbol indicates that two calculations are necessary, one using $+$ and one using $-$.) Here's a sample equation:

$$3x^2 - 2x = 3x + 2$$

First, the quadratic expression should be set equal to zero:

$$3x^2 - 5x - 2 = 0$$

Then the values are substituted for $a(3)$, $b(-5)$, and $c(-2)$ into the formula:

$$x = \frac{-(-5) \pm \sqrt{(-5)^2 - 4(3)(-2)}}{2(3)}$$

Simplification yields:

$$x = \frac{5 \pm \sqrt{49}}{6} \rightarrow x = \frac{5 \pm 7}{6}$$

Calculating two values for x using $+$ and $-$ yields:

$$x = \frac{5 + 7}{6}; x = \frac{5 - 7}{6}$$

Simplification yields:

$$x = 2 \text{ or } -\frac{1}{3}.$$

Just as with any equation, solutions should be checked by substituting the value into the original equation.

Adding, Subtracting, and Multiplying Polynomial Expressions

A polynomial expression is a monomial (one term) or the sum of monomials (more than one term separated by addition or subtraction). A polynomial in standard form consists of terms with variables written in descending exponential order and with any like terms combined.

Adding/Subtracting Polynomials

When adding or subtracting polynomials, each polynomial should be written in parenthesis; the negative sign should be distributed when necessary, and like terms need to be combined. Here's a sample equation: add $3x^3 + 4x - 3$ to $x^3 - 3x^2 + 2x - 2$. The sum is set as follows:

$$(x^3 - 3x^3 + 2x - 2) + (3x^3 + 4x - 3)$$

In front of each set of parentheses is an implied positive 1, which, when distributed, does not change any of the terms.

Therefore, the parentheses should be dropped and like terms should be combined:

$$x^3 - 3x^2 + 2x - 2 + 3x^3 + 4x - 3$$

$$4x^3 - 3x^2 + 6x - 5$$

Here's another sample equation: subtract $3x^3 + 4x - 3$ from $x^3 - 3x^2 + 2x - 2$. The difference should be set as follows:

$$(x^3 - 3x^2 + 2x - 2) - (3x^3 + 4x - 3)$$

The implied $+1$ in front of the first set of parentheses will not change those four terms; however, distributing the implied -1 in front of the second set of parentheses will change the sign of each of those three terms:

$$x^3 - 3x^2 + 2x - 2 - 3x^3 - 4x + 3$$

Combining like terms yields:

$$-2x^3 - 3x^2 - 2x + 1$$

Multiplying Polynomials

When multiplying monomials, the coefficients are multiplied and exponents of the same variable are added. For example:

$$-5x^3y^2z \times 2x^2y^5z^3 = -10x^5y^7z^4$$

When multiplying polynomials, the monomials should be distributed and multiplied, then any like terms should be combined and written in standard form. Here's a sample equation:

$$2x^3(3x^2 + 2x - 4)$$

First, $2x^3$ should be multiplied by each of the three terms in parentheses:

$$2x^3 \times 3x^2 + 2x^3 \times 2x + 2x^3 \times -4 = 6x^5 + 4x^4 - 8x^3$$

Multiplying binomials will sometimes be taught using the FOIL method (where the products of the first, outside, inside, and last terms are added together). However, it may be easier and more consistent to think of it in terms of distributing. Both terms of the first binomial should be distributed to both terms of the second binomial. For example, the product of binomials $(2x + 3)(x - 4)$ can be calculated by distributing $2x$ and distributing 3:

$$2x \times x + 2x \times -4 + 3 \times x + 3 \times -4$$

$$2x^2 - 8x + 3x - 12$$

Combining like terms yields $2x^2 - 5x - 12$.

The general principle of distributing each term can be applied when multiplying polynomials of any size. To multiply $(x^2 + 3x - 1)(5x^3 - 2x^2 + 2x + 3)$, all three terms should be distributed from the first polynomial to each of the four terms in the second polynomial and then any like terms should be combined. If a problem requires multiplying more than two polynomials, two at a time can be multiplied and combined until all have been multiplied. To multiply $(x + 3)(2x - 1)(x + 5)$, two polynomials should be chosen to multiply together first. Multiplying the last two results in:

$$(2x - 1)(x + 5) = 2x^2 + 9x - 5$$

That product should then be multiplied by the third polynomial:

$$(x + 3)(2x^2 + 9x - 5)$$

The final answer should equal:

$$2x^3 + 15x^2 + 36x - 15$$

Solving an Equation in One Variable that Contains Radicals or Contains the Variable in the Denominator of a Fraction

Equations with radicals containing numbers only as the radicand are solved the same way that an equation without a radical would be. For example, $3x + \sqrt{81} = 45$ would be solved using the same steps as if solving $2x + 4 = 12$. Radical equations are those in which the variable is part of the radicand. For example, $\sqrt{5x + 1} - 6 = 0$ and $\sqrt{x - 3} + 5 = x$ would be considered radical equations.

Radical Equations

To solve a radical equation, the radical should be isolated and both sides of the equation should be raised to the same power to cancel the radical. Raising both sides to the second power will cancel a square root, raising to the third power will cancel a cube root, etc. To solve $\sqrt{5x + 1} - 6 = 0$, the radical should be isolated first: $\sqrt{5x + 1} = 6$. Then both sides should be raised to the second power:

$$\left(\sqrt{5x + 1}\right)^2 = (6)^2 \to 5x + 1 = 36$$

Lastly, the linear equation should be solved: $x = 7$.

Radical Equations with Extraneous Solutions

If a radical equation contains a variable in the radicand and a variable outside of the radicand, it must be checked for extraneous solutions. An extraneous solution is one obtained by following the proper process for solving an equation but does not "check out" when substituted into the original equation.

Here's a sample equation: $\sqrt{x-3} + 5 = x$. Isolating the radical yields $\sqrt{x-3} = x - 5$. Next, both sides should be squared to cancel the radical:

$$(\sqrt{x-3})^2 = (x-5)^2 \rightarrow x - 3 = (x-5)(x-5)$$

The binomials should be multiplied: $x - 3 = x^2 - 10x + 25$. The quadratic equation is then solved:

$$0 = x^2 - 11x + 28$$

$$0 = (x - 7)(x - 4)$$

$$x - 7 = 0; x - 4 = 0$$

$$x = 7 \text{ or } x = 4$$

To check for extraneous solutions, each answer can be substituted, one at a time, into the original equation. Substituting 7 for x, results in $7 = 7$. Therefore, 7 is a solution. Substituting 4 for x results in $6 = 4$. This is false; therefore, 4 is an extraneous solution.

Equations with a Variable in the Denominator of a Fraction

For equations with variables in the denominator, if the equation contains two rational expressions (on opposite sides of the equation, or on the same side and equal to zero), it can be solved like a proportion. Here's an equation to consider: $\frac{5}{2x-2} = \frac{15}{x^2-1}$. First, cross-multiplying yields:

$$5(x^2 - 1) = 15(2x - 2)$$

Distributing yields:

$$5x^2 - 5 = 30x - 30$$

In solving the quadratic equation, it is determined that $x = 1$ or $x = 5$. Solutions must be checked to see if they are extraneous. Extraneous solutions either produce a false statement when substituted into the original equation or create a rational expression with a denominator of zero (dividing by zero is undefined). Substituting 5 into the original equation produces $\frac{5}{8} = \frac{5}{8}$; therefore, 5 is a solution. Substituting 1 into the original equation results in both denominators equal to zero; therefore, 1 is an extraneous solution.

If an equation contains three or more rational expressions: the least common denominator (LCD) needs to be found for all the expressions, then both sides of the equation should be multiplied by the LCD. The LCD consists of the lowest number that all coefficients divide evenly into and for every variable, the highest power of that variable. Here's a sample equation: $\frac{3}{5x} - \frac{4}{3x} = \frac{1}{3}$. The LCD would be $15x$. Both sides of the equation should be multiplied by $15x$:

$$15x\left(\frac{3}{5x} - \frac{4}{3x}\right) = 15x\left(\frac{1}{3}\right) \rightarrow \frac{45x}{5x} - \frac{60x}{3x}$$

$$\frac{15x}{3} \rightarrow 9 - 20 = 5x \rightarrow x = -2\frac{1}{2}$$

Any extraneous solutions should be identified.

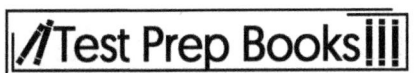

Solving a System of One Linear Equation and One Quadratic Equation

A system of equations consists of two variables in two equations. A solution to the system is an ordered pair (x, y) that makes both equations true. When displayed graphically, a solution to a system is a point of intersection between the graphs of the equations. When a system consists of one linear equation and one quadratic equation, there may be one, two, or no solutions. If the line and parabola intersect at two points, there are two solutions to the system; if they intersect at one point, there is one solution; if they do not intersect, there is no solution.

Systems with One Linear Equation and One Quadratic Equation

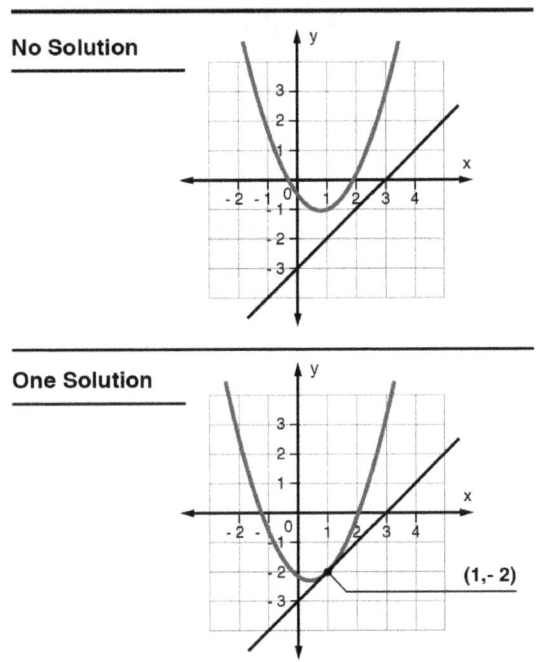

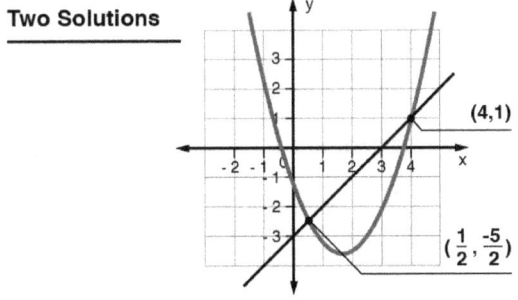

One method for solving a system of one linear equation and one quadratic equation is to graph both functions and identify point(s) of intersection. This, however, is not always practical. Graph paper may not be available or the intersection points may not be easily identified. Solving the system algebraically involves using the substitution method. Consider the following system: $y = x^2 + 9x + 11; y = 2x - 1$.

The equivalent value of y should be substituted from the linear equation $(2x - 1)$ into the quadratic equation. The resulting equation is

$$2x - 1 = x^2 + 9x + 11$$

Next, this quadratic equation should be solved using the appropriate method: factoring, taking square roots, or using the quadratic formula. Solving this quadratic equation by factoring results in $x = -4$ or $x = -3$. Next, the corresponding y-values should be found by substituting the x-values into the original linear equation:

$$y = 2(-4) - 1; y = 2(-3) - 1$$

The solutions should be written as ordered pairs: (-4, -9) and (-3, -7). Finally, the possible solutions should be checked by substituting each into both of the original equations. In this case, both solutions "check out."

Rewriting Simple Rational Expressions

A rational expression is an algebraic expression including variables that look like a fraction. In simplest form, the numerator and denominator of a rational expression do not have common divisors (factors). To simplify a rational expression, the numerator and denominator should be factored; then any common factors in the numerator and denominator should be canceled. To simplify the foll, the numerator and denominator should be written as a product of its factors:

$$\frac{3x^2y}{12xy^3}$$

$$\frac{3 \cdot x \cdot x \cdot y}{2 \cdot 2 \cdot 3 \cdot x \cdot y \cdot y \cdot y}$$

Canceling common factors leaves: $\frac{x}{2 \cdot 2 \cdot y \cdot y}$. Multiplying the remaining factors results in $\frac{x}{4y^2}$.

Here's a rational expression:

$$\frac{x^2 - 1}{x^2 - x - 2}$$

Factoring the numerator and denominator produces:

$$\frac{(x + 1)(x - 1)}{(X - 2)(x + 1)}$$

Each binomial in parentheses is a factor and only the exact same binomial would cancel that factor. By canceling factors, the expression is simplified to:

$$\frac{x - 1}{x - 2}$$

The variable x itself is not a factor. Therefore, they do not cancel each other out.

Multiplying/Dividing Rational Expressions

When multiplying or dividing rational expressions, the basic concepts of operations with fractions are used. To multiply, (1) all numerators and denominators need to be factored, (2) common factors should be canceled between any numerator and any denominator, (3) the remaining factors of the numerator and the remaining factors of the denominator should be multiplied, and (4) the expression should be checked to see whether it can be simplified further.

To multiply the following, each numerator and denominator should be written as a product of its factors:

$$\frac{4a^4}{3} \times \frac{6}{5a^2}$$

$$\frac{2 \cdot 2 \cdot a \cdot a \cdot a \cdot a}{3} \times \frac{3 \cdot 2}{5 \cdot a \cdot a}$$

After canceling common factors, the remaining expression is:

$$\frac{2 \cdot 2 \cdot a \cdot a}{1} \times \frac{2}{5}$$

A factor of 1 remains if all others are canceled. Multiplying remaining factors produces:

$$\frac{8a^2}{5}$$

To divide rational expressions, the expression should be changed to multiplying by the reciprocal of the divisor (just as with fractions: $\frac{1}{2} \div \frac{3}{4} = \frac{1}{2} \times \frac{4}{3}$); then follow the process for multiplying rational expressions.

Here's a sample expression:

$$\frac{2x}{x^2 - 16} \div \frac{4x^2 + 6x}{x^2 + 6x + 8}$$

First, the division problem should be changed to a multiplication problem:

$$\frac{2x}{x^2 - 16} \times \frac{x^2 + 6x + 8}{4x^2 + 6x}$$

Then, the equation should be factored:

$$\frac{2x}{(x + 4)(x - 4)} \times \frac{(x + 4)(x + 2)}{2x(2x + 3)}$$

Canceling yields:

$$\frac{1}{(x - 4)} \times \frac{(x + 2)}{(2x + 3)}$$

Multiplying the remaining factors produces:

$$\frac{x+2}{2x^2 - 5x - 12}$$

Adding/Subtracting Rational Expressions

Just as with adding and subtracting fractions, to add or subtract rational expressions, a common denominator is needed. (The numerator is added or subtracted and the denominator stays the same.) If the expressions have like denominators, subtraction should be changed to add the opposite (a -1 is distributed to each term in the numerator of the expression being subtracted); the denominators should be factored and the expressions added; the numerator should then be factored; and the equation should be simplified if possible. Here's a sample expression:

$$\frac{2x^2 + 4x - 3}{x + 3} - \frac{x^2 - 2x - 12}{x + 3}$$

Changing subtraction to add the opposite yields:

$$\frac{2x^2 + 4x - 3}{x + 3} + \frac{-x^2 + 2x + 12}{x + 3}$$

The denominator cannot be factored, so the expression should be added, resulting in:

$$\frac{x^2 + 6x + 9}{x + 3}$$

Simplification is performed by factoring the numerator: $\frac{(x+3)(x+3)}{(x+3)}$. Canceling yields: $\frac{x+3}{1}$, or simply x + 3.

To add or subtract rational expressions with unlike denominators, the denominators must be changed by finding the least common multiple (LCM) of the expressions. To find the LCM, each expression should be factored and the product should be formed using each factor the greatest number of times it occurs. The LCM of $12xy^2$ and $15x^3y$ would be $60x^3y^2$. The LCM of $x^2 + 5x + 4$ (which factors to $(x+4)(x+1)$) and $x^2 + 2x + 1$ (which factors to $(x+1)(x+1)$) would be $(x+4)(x+1)(x+1)$.

To add or subtract expressions with unlike denominators: (1) subtraction should be changed to add the opposite; (2) the denominators are factored; (3) an LCM should be determined for the denominators; (4) the numerator and denominator of each expression should be multiplied by the missing factor(s); (5) the expressions that now have like denominators should be added; (6) the numerator should be factored; and (7) simplification should be performed if possible. Here's a sample expression:

$$\frac{x^2 + 6x + 11}{x^2 + 7x + 12} - \frac{2}{x + 3}$$

First, subtraction should be changed to addition:

$$\frac{x^2 + 6x + 11}{x^2 + 7x + 12} + \frac{-2}{x + 3}$$

Then, the denominators are factored:

$$\frac{x^2 + 6x + 11}{(x + 4)(x + 3)} + \frac{-2}{x + 3}$$

The LCM of $(x + 4)(x + 3)$ and $(x + 3)$ should be determined, which is $(x + 4)(x + 3)$. The numerator and denominator should be multiplied by the missing factor:

$$\frac{x^2 + 6x + 11}{(x + 4)(x + 3)} + \frac{-2}{x + 3} \times \frac{(x + 4)}{(x + 4)}$$

$$\frac{x^2 + 6x + 11}{(x + 4)(x + 3)} + \frac{-2x - 8}{(x + 4)(x + 3)}$$

The expressions should be added, resulting in:

$$\frac{x^2 + 4x + 3}{(x + 4)(x + 3)}$$

The numerator should be factored:

$$\frac{(x + 3)(x + 1)}{(x + 4)(x + 3)}$$

Simplifying yields:

$$\frac{x + 1}{x + 4}$$

Interpreting Parts of Nonlinear Expressions in Terms of Their Context

When a nonlinear function is used to model a real-life scenario, some aspects of the function may be relevant while others may not. The context of each scenario will dictate what should be used. In general, x- and y-intercepts will be points of interest. A y-intercept is the value of y when x = zero; and an x-intercept is the value of x when y = zero. Suppose a nonlinear function models the value of an investment(y) over the course of time(x). It would be relevant to determine the initial value (the y-intercept where time = zero), as well as any point in time in which the value would be zero (the x-intercept).

Another aspect of a function that is typically desired is the rate of change. This tells how fast the outputs are growing or decaying with respect to given inputs. For polynomial functions, the rate of change can be estimated by the highest power of the function. Polynomial functions also include absolute and/or relative minimums and maximums. Functions modeling production or expenses should be considered. Maximum and minimum values would be relevant aspects of these models.

Finally, the domain and range for a function should be considered for relevance. The domain consists of all input values and the range consists of all output values. For instance, a function could model the volume of a container to be produced in relation to its height. Although the function that models the scenario may include negative values for inputs and outputs, these parts of the function would obviously not be relevant.

Understanding the Relationship Between Zeros and Factors of Polynomials

The zeros of a function are the *x*-intercepts of its graph. They are called zeros because they are the *x*-values for which y = 0.

Finding Zeros

To find the zeros of a polynomial function, it should be written in factored form, then each factor should be set equal to zero and solved. To find the zeros of the function $y = 3x^3 - 3x^2 - 36x$, the polynomial should be factored first. Factoring out a GCF results in $y = 3x(x^2 - x - 12)$. Then factoring the quadratic function yields:

$$y = 3x(x - 4)(x + 3)$$

Next, each factor should be set equal to zero: $3x = 0; x - 4 = 0; x + 3 = 0$. By solving each equation, it is determined that the function has zeros, or *x*-intercepts, at 0, 4, and -3.

Writing a Polynomial with Given Zeros

Given zeros for a polynomial function, to write the function, a linear factor corresponding to each zero should be written. The linear factor will be the opposite value of the zero added to *x*. Then the factors should be multiplied and the function written in standard form. To write a polynomial with zeros at -2, 3, and 3, three linear factors should be written:

$$y = (x + 2)(x - 3)(x - 3)$$

Then, multiplication is used to convert the equation to standard form, producing:

$$y = x^3 - 4x^2 - 3x + 18$$

Dividing Polynomials by Linear Factors

To determine if a linear binomial is a factor of a polynomial, the polynomial should be divided by the binomial. If there is no remainder (it divides evenly), then the binomial is a factor of the polynomial. To determine if a value is a zero of a function, a binomial can be written from that zero and tested by division. To divide a polynomial by a linear factor, the terms of the dividend should be divided by the linear term of the divisor; the same process as long division of numbers (divide, multiply, subtract, drop down, and repeat) should be followed.

$$\frac{divisor\sqrt{quotient}}{dividend}$$

Remember that when subtracting a binomial, the signs of both terms should be changed. Here's a sample equation: divide $9x^3 - 18x^2 - x + 2$ by $3x + 1$. First, the problem should be set up as long division:

$$3x + 1 \overline{) 9x^3 - 18x^2 - x + 2}$$

Then the first term of the dividend ($9x^3$) should be divided by the linear term of the divisor ($3x$):

$$3x + 1 \overline{) \begin{matrix} 3x^2 \\ 9x^3 - 18x^2 - x + 2 \end{matrix}}$$

Next, the divisor should be multiplied by that term of the quotient:

$$\begin{array}{r} 3x^2 - 7x + 2 \\ 3x+1 \overline{\smash{)}\, 9x^3 - 18x^2 - x + 2} \\ -9x^3 - 3x^2 \end{array}$$

Subtraction should come next:

$$\begin{array}{r} 3x^2 - 7x + 2 \\ 3x+1 \overline{\smash{)}\, 9x^3 - 18x^2 - x + 2} \\ \underline{-9x^3 - 3x^2 } \\ -21x^2 \end{array}$$

Now, the next term (-x) should be dropped down:

$$\begin{array}{r} 3x^2 - 7x + 2 \\ 3x+1 \overline{\smash{)}\, 9x^3 - 18x^2 - x + 2} \\ \underline{-9x^3 - 3x^2 } \\ -21x^2 - x \end{array}$$

Then the process should be repeated, dividing $-21x^2$ by $3x$:

$$\begin{array}{r} 3x^2 - 7x + 2 \\ 3x+1 \overline{\smash{)}\, 9x^3 - 18x^2 - x + 2} \\ \underline{-9x^3 - 3x^2 } \\ -21x^2 - x \\ \underline{+21x^2 + 7x } \\ 6x \end{array}$$

The next term (2) should be dropped and repeated by dividing $6x$ by $3x$:

$$\begin{array}{r} 3x^2 - 7x + 2 \\ 3x+1 \overline{\smash{)}\, 9x^3 - 18x^2 - x + 2} \\ \underline{-9x^3 - 3x^2 } \\ -21x^2 - x \\ \underline{+21x^2 + 7x } \\ 6x + 2 \\ \underline{-6x - 2} \\ 0 \end{array}$$

There is no remainder; therefore, $3x + 1$ is a factor of:

$$9x^3 - 18x^2 - x + 2$$

By the definition of factors:

$$(3x + 1)(3x^2 - 7x + 2) = 9x^3 - 18x^2 - x + 2$$

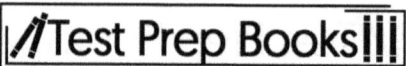

The quadratic expression can further be factored to produce:

$$(3x + 1)(3x - 1)(x - 2)$$

Understanding a Nonlinear Relationship Between Two Variables

Questions on this material will assess the ability of test takers to make connections between linear or nonlinear equations and their graphical representations. It will also require interpreting graphs in relation to systems of equations.

Graphs of Polynomial Functions

A polynomial function consists of a monomial or sum of monomials arranged in descending exponential order. The graph of a polynomial function is a smooth continuous curve that extends infinitely on both ends. From the equation of a polynomial function, the following can be determined: (1) the end behavior of the graph—does it rise or fall to the left and to the right; (2) the y-intercept and x-intercept(s) and whether the graph simply touches or passes through each x-intercept; and (3) the largest possible number of turning points, where the curve changes from rising to falling or vice versa. To graph the function, these three aspects of the graph should be determined and extra points between the intercepts should be found if necessary.

End Behavior

The end behavior of the graph of a polynomial function can be determined by the degree of the function (largest exponent) and the leading coefficient (coefficient of the term with the largest exponent). There are four possible scenarios for the end behavior: (1) if the degree is odd and the coefficient is positive, the graph falls to the left and rises to the right; (2) if the degree is odd and the coefficient is negative, the graph rises to the left and falls to the right; (3) if the degree is even and the coefficient is positive, the graph rises to the left and rises to the right, or (4) if the degree is even and the coefficient is negative, the graph falls to the left and falls to the right.

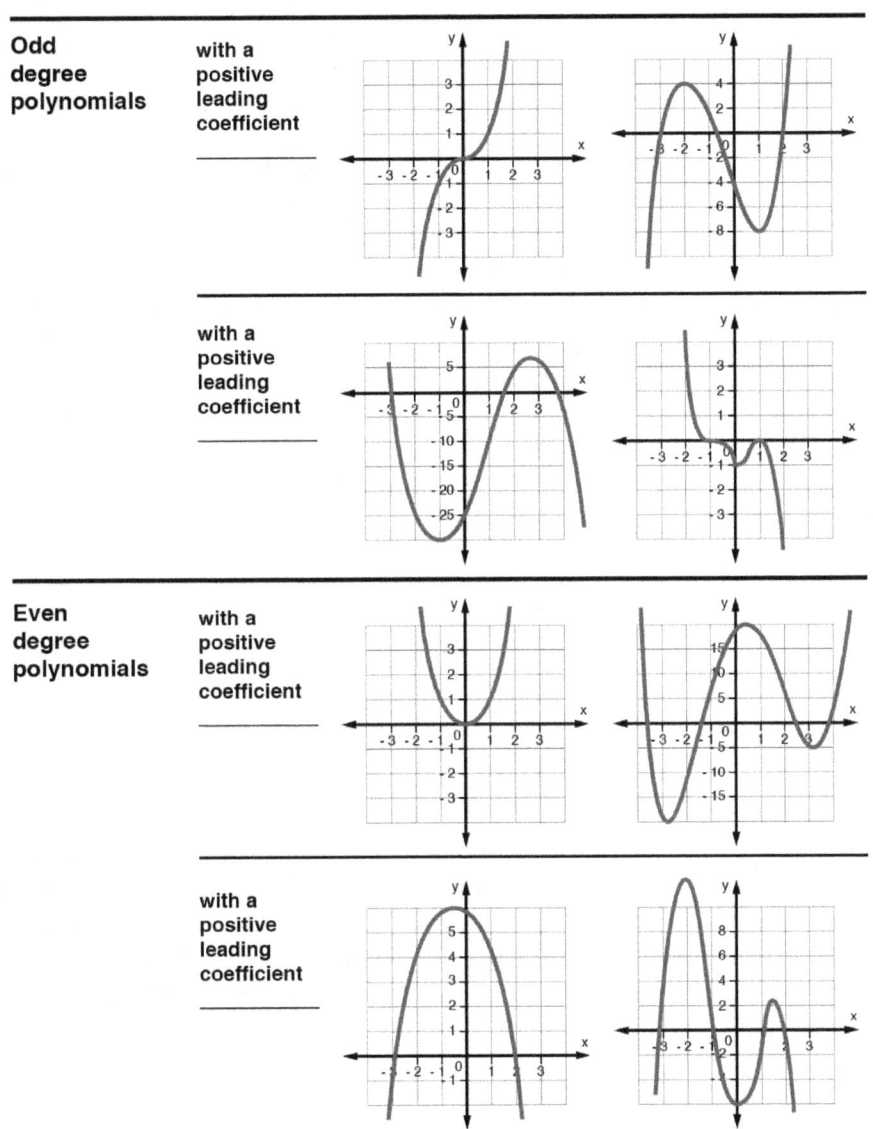

X and Y-Intercepts

The y-intercept for any function is the point at which the graph crosses the y-axis. At this point $x = 0$; therefore, to determine the y-intercept, $x = 0$ should be substituted into the function and solved for y. For a given zero of a function, the graph can either pass through that point or simply touch that point (the graph turns at that zero). This is determined by the multiplicity of that zero. The multiplicity of a zero is the number of times its corresponding factor is multiplied to obtain the function in standard form. For example, $y = x^3 - 4x^2 - 3x + 18$ can be written in factored form as $y = (x + 2)(x - 3)(x - 3)$ or $y = (x + 2)(x - 3)^2$. The zeros of the function would be -2 and 3. The zero at -2 would have a multiplicity of 1, and the zero at 3 would have a multiplicity of 2. If a zero has an even multiplicity, then the graph touches the x-axis at that zero and turns around. If a zero has an odd multiplicity, then the graph crosses the x-axis at that zero.

Turning Points

The graph of a polynomial function can have, at most, a number of turning points equal to one less than the degree of the function. It is possible to have fewer turning points than this value. For example, the function $y = 3x^5 + 2x^2 - 3x$ could have no more than four turning points.

Using Function Notation, and Interpreting Statements Using Function Notation.

Function notation is covered in the *Function/Linear Equation Notation* section under *Heart of Algebra*.

Addition, Subtraction, Multiplication and Division of Functions

Functions denoted by *f(x)*, *g(x)*, etc., can be added, subtracted, multiplied, or divided. For example, the function $f(x) = 15x + 100$ represents the cost to have a catered party at a banquet hall (where *x* represents the number of guests); and the function $g(x) = 10x$ represents the cost for unlimited drinks at the party. The total cost of a catered party with unlimited drinks can be represented by adding the functions *f(x)* and *g(x)*. In this case $f(x) + g(x) = (15x + 100) + (10x)$; therefore, $f(x) + g(x) = 25x + 100$. $(f(x) + g(x))$ can also be written $(f + g)(x)$. To add, subtract, multiply, or divide functions, the values of the functions should be substituted and the rules for operations with polynomials should be followed. It should be noted:

$$(f - g)(x) = f(x) - g(x); (f \times g)(x) = f(x) \times g(x)$$

and

$$\left(\frac{f}{g}\right)(x) = \frac{f(x)}{g(x)}$$

Composition of Functions

A composite function is one in which two functions are combined such that the output from the first function becomes the input for the second function (one function should be applied after another function). The composition of a function written as $(g \circ f)(x)$ or $g(f(x))$ is read "g of f of x." The inner function, *f(x)*, would be evaluated first and the answer would be used as the input of the outer function, *g(x)*. To determine the value of a composite function, the value of the inner function should be substituted for the variable of the outer function.

Here's a sample problem:

A store is offering a 20% discount on all of its merchandise. You have a coupon worth $5 off any item.

The cost of an item with the 20% discount can be modeled by the function: $d(x) = 0.8x$. The cost of an item with the coupon can be modeled by the function $c(x) = x - 5$. A composition of functions to model the cost of an item applying the discount first and then the coupon would be $c(d(x))$.

Replacing $d(x)$ with its value ($0.8x$) results in $c(0.8x)$. By evaluating the function $c(x)$ with an input of $0.8x$, it is determined that $c(d(x)) = 0.8x - 5$. To model the cost of an item if the coupon is applied first and then the discount, $d(c(x))$ should be determined. The result would be: $d(c(x)) = 0.8x - 4$.

Evaluating Functions

If a problem asks to evaluate with operations between functions, the new function should be determined and then the given value should be substituted as the input of the new function. To find $(f \times g)(3)$ given $f(x) = x + 1$ and $g(x) = 2x - 3$, the following should be determined:

$$(f \times g)(x) = f(x) \times g(x) = (x + 1)(2x - 3) = 2x^2 - x - 3$$

Therefore, $(f \times g)(x) = 2x^2 - x - 3$.

To find $(f \times g)(3)$, the function $(f \times g)(x)$ needs to be evaluated for an input of 3:

$$(f \times g)(3) = 2(3)^2 - (3) - 3 = 12$$

Therefore, $(f \times g)(3) = 12$.

Using Structure to Isolate or Identify a Quantity of Interest

Formulas are mathematical expressions that define the value of one quantity given the value of one or more different quantities. A formula or equation expressed in terms of one variable can be manipulated to express the relationship in terms of any other variable. The equation $y = 3x + 2$ is expressed in terms of the variable y. By manipulating the equation, it can be written as $x = \frac{y-2}{3}$, which is expressed in terms of the variable x. To manipulate an equation or formula to solve for a variable of interest, how the equation would be solved if all other variables were numbers should be considered. The same steps for solving should be followed, leaving operations in terms of the variables, instead of calculating numerical values.

The formula $P = 2l + 2w$ expresses how to calculate the perimeter of a rectangle given its length and width. To write a formula to calculate the width of a rectangle given its length and perimeter, the previous formula relating the three variables should be used and the variable w should be solved. If P and l were numerical values, this would be a two-step linear equation solved by subtraction and division. To solve the equation $P = 2l + 2w$ for w, $2l$ should be subtracted from both sides: $P - 2l = 2w$. Then both sides should be divided by 2: $\frac{P - 2l}{2} = w$ or $\frac{P}{2} - l = w$.

The distance formula between two points on a coordinate plane can be found using the formula:

$$d = \sqrt{(x_2 - x_1)^2 + (y_2 - y_1)^2}$$

A problem might require determining the x-coordinate of one point (x_2), given its y-coordinate (y_2) and the distance (d) between that point and another given point (x_1, y_1). To do so, the above formula for x_1 should be solved just as a radical equation containing numerical values in place of the other variables. Both sides should be squared; the quantity should be subtracted $(y_2 - y_1)^2$; the square root of both sides should be taken; x_1 should be subtracted to produce:

$$\sqrt{d^2 - (y_2 - y_1)^2} + x_1 = x_2$$

PSAT 8/9 Math Practice Test #1

1. Which of the following inequalities is equivalent to $3 - \frac{1}{2}x \geq 2$?
   a. $x \geq 2$
   b. $x \leq 2$
   c. $x \geq 1$
   d. $x \leq 1$

2. If $g(x) = x^3 - 3x^2 - 2x + 6$ and $f(x) = 2$, then what is $g(f(x))$?
   a. -26
   b. 6
   c. $2x^3 - 6x^2 - 4x + 12$
   d. -2

3. The graph of which function has an x-intercept of -2?
   a. $y = 2x - 3$
   b. $y = 4x + 2$
   c. $y = x^2 + 5x + 6$
   d. $y = -\frac{1}{2} \times 2^x$

4. The table below displays the number of three-year-olds at Kids First Daycare who are potty-trained and those who still wear diapers.

	Potty-trained	Wear diapers	
Boys	26	22	48
Girls	34	18	52
	60	40	

What is the probability that a three-year-old girl chosen at random from the school is potty-trained?
   a. 52 percent
   b. 34 percent
   c. 65 percent
   d. 57 percent

5. What is the solution to the following system of equations?
$$x^2 - 2x + y = 8$$
$$x - y = -2$$
   a. $(-2, 3)$
   b. There is no solution.
   c. $(-2, 0)$ $(1, 3)$
   d. $(-2, 0)$ $(3, 5)$

6. An equation for the line passing through the origin and the point (2, 1) is
   a. $y = 2x$
   b. $y = \frac{1}{2}x$
   c. $y = x - 2$
   d. $y = x - 2$

7. What type of function is modeled by the values in the following table?

X	f(x)
1	2
2	4
3	8
4	16
5	32

    a. Linear
    b. Exponential
    c. Quadratic
    d. Cubic

8. Two cards are drawn from a shuffled deck of 52 cards. What's the probability that both cards are Kings if the first card isn't replaced after it's drawn?
    a. $\frac{1}{169}$
    b. $\frac{1}{221}$
    c. $\frac{1}{13}$
    d. $\frac{4}{13}$

9. Write the expression for six less than three times the sum of twice a number and one.
    a. $2x + 1 - 6$
    b. $3x + 1 - 6$
    c. $3(x + 1) - 6$
    d. $3(2x + 1) - 6$

10. $(2x - 4y)^2 =$
    a. $4x^2 - 16xy + 16y^2$
    b. $4x^2 - 8xy + 16y^2$
    c. $4x^2 - 16xy - 16y^2$
    d. $2x^2 - 8xy + 8y^2$

11. What are the zeros of $f(x) = x^2 + 4$?
    a. $x = -4$
    b. $x = \pm 2i$
    c. $x = \pm 2$
    d. $x = \pm 4i$

12. Which of the following shows the correct result of simplifying the following expression:
$$(7n + 3n^3 + 3) + (8n + 5n^3 + 2n^4)$$
    a. $9n^4 + 15n - 2$
    b. $2n^4 + 5n^3 + 15n - 2$
    c. $9n^4 + 8n^3 + 15n$
    d. $2n^4 + 8n^3 + 15n + 3$

13. What is the slope of this line?

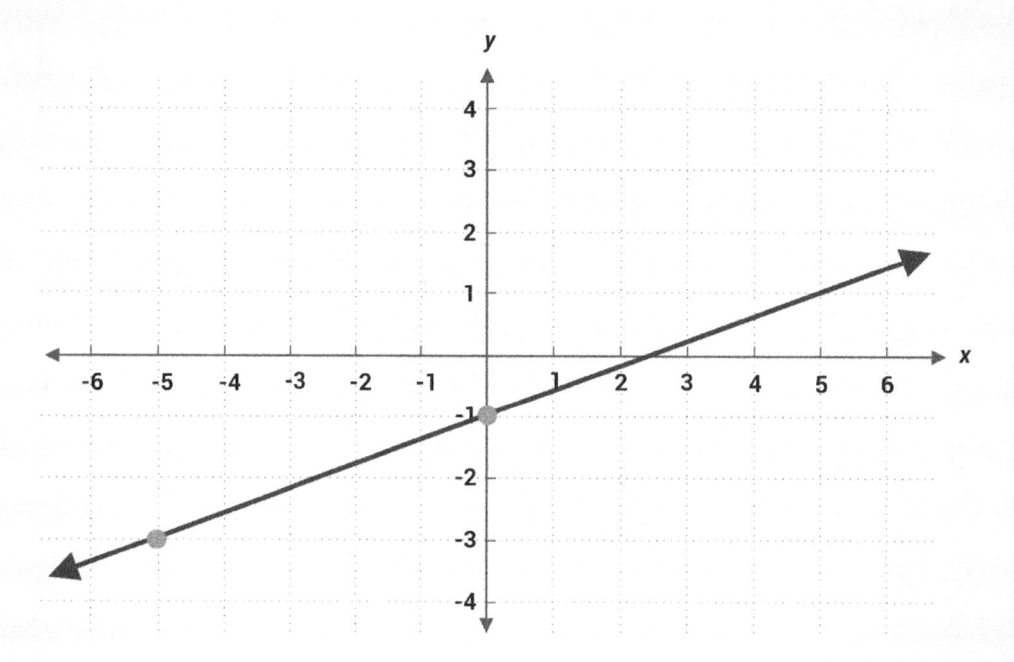

a. 2
b. $\frac{5}{2}$
c. $\frac{1}{2}$
d. $\frac{2}{5}$

14. What is the product of the following expression?
$$(4x - 8)(5x^2 + x + 6)$$
a. $20x^3 - 36x^2 + 16x - 48$
b. $6x^3 - 41x^2 + 12x + 15$
c. $204 + 11x^2 - 37x - 12$
d. $2x^3 - 11x^2 - 32x + 20$

15. What is the solution for the following equation?
$$\frac{x^2 + x - 30}{x - 5} = 11$$
a. $x = -6$
b. There is no solution.
c. $x = 16$
d. $x = 5$

16. If x is not zero, then $\frac{3}{x} + \frac{5u}{2x} - \frac{u}{4} =$

a. $\frac{12+10u-ux}{4x}$

b. $\frac{3+5u-ux}{x}$

c. $\frac{12x+10u+ux}{4x}$

d. $\frac{12+10u-u}{4x}$

17. What are the zeros of the function: $f(x) = x^3 + 4x^2 + 4x$?
   a. -2
   b. 0, -2
   c. 2
   d. 0, 2

18. Is the following function even, odd, neither, or both?
$$y = \frac{1}{2}x^4 + 2x^2 - 6$$
   a. Even
   b. Odd
   c. Neither
   d. Both

19. A pair of dice is thrown, and the sum of the two scores is calculated. What's the expected value of the roll?
   a. 5
   b. 6
   c. 7
   d. 8

20. What's the probability of rolling a 6 at least once in two rolls of a die?
   a. $\frac{1}{3}$
   b. $\frac{1}{36}$
   c. $\frac{1}{6}$
   d. $\frac{11}{36}$

21. Karen gets paid a weekly salary and a commission for every sale that she makes. The table below shows the number of sales and her pay for different weeks.

Sales	2	7	4	8
Pay	$380	$580	$460	$620

Which of the following equations represents Karen's weekly pay?
   a. $y = 90x + 200$
   b. $y = 90x - 200$
   c. $y = 40x + 300$
   d. $y = 40x - 300$

22. The graph shows the position of a car over a 10-second time interval. Which of the following is the correct interpretation of the graph for the interval 1 to 3 seconds?

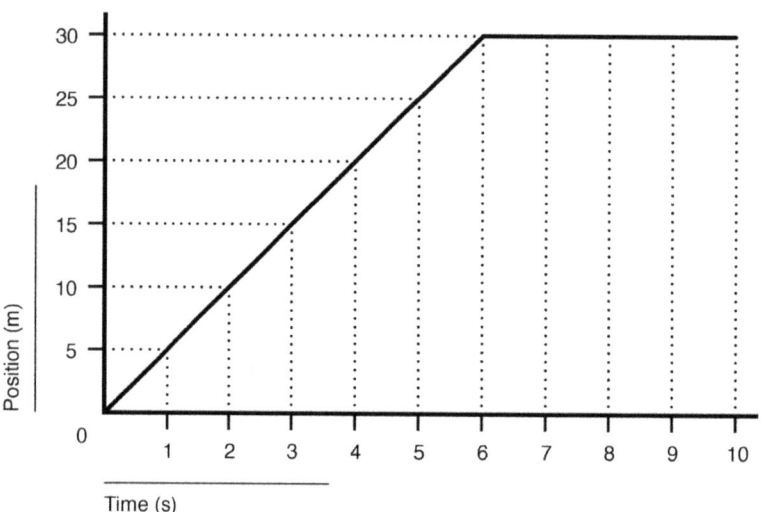

a. The car remains in the same position.
b. The car is traveling at a speed of 5m/s.
c. The car is traveling up a hill.
d. The car is traveling at 5 mph.

23. Which of the ordered pairs below is a solution to the following system of inequalities?
$$y > 2x - 3$$
$$y < -4x + 8$$

a. (4, 5)
b. (−3, −2)
c. (3, −1)
d. (5, 2)

24. Which equation best represents the scatterplot below?

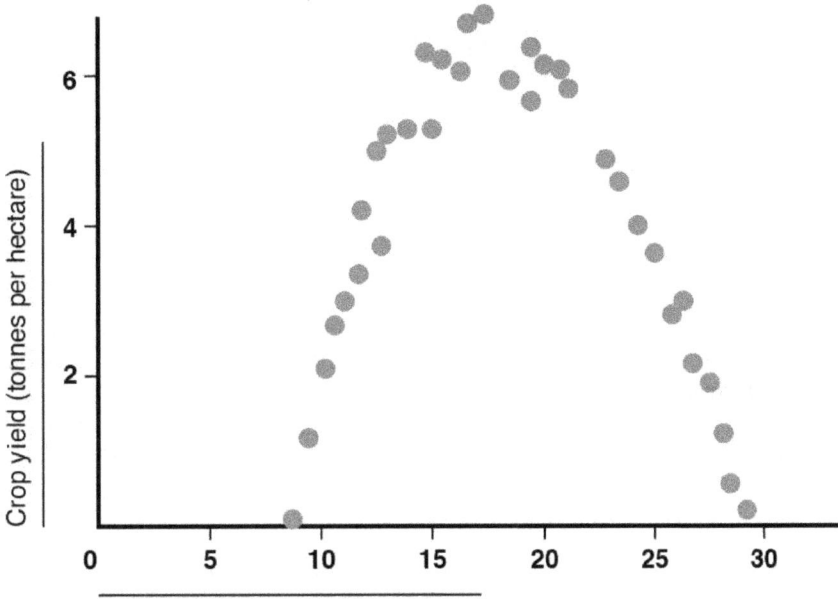

a. $y = 3x - 4$
b. $y = 2x^2 + 7x - 9$
c. $y = (3)(4^x)$
d. $y = -\frac{1}{14}x^2 + 2x - 8$

25. Suppose an investor deposits $1,200 into a bank account that accrues 1 percent interest per month. Assuming x represents the number of months since the deposit and y represents the money in the account, which of the following exponential functions models the scenario?
a. $y = (0.01)(1200^x)$
b. $y = (1200)(0.01^x)$
c. $y = (1.01)(1200^x)$
d. $y = (1200)(1.01^x)$

26. Which of the following is the result of simplifying the expression:
$$\frac{4a^{-1}b^3}{a^4b^{-2}} * \frac{3a}{b}$$

a. $12a^3b^5$

b. $12\frac{b^4}{a^4}$

c. $\frac{12}{a^4}$

d. $7\frac{b^4}{a}$

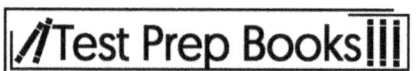

27. What is the simplified quotient of the following equation?

$$\frac{5x^3}{3x^2y} \div \frac{25}{3y^9}$$

a. $\frac{125x}{9y^{10}}$

b. $\frac{x}{5y^8}$

c. $\frac{5}{xy^8}$

d. $\frac{xy^8}{5}$

No Calculator Questions

28. Five of six numbers have a sum of 25. The average of all six numbers is 6. What is the sixth number?
    a. 8
    b. 10
    c. 11
    d. 12

29. Solve for x: $\frac{2x}{5} - 1 = 59$.
    a. 60
    b. 145
    c. 150
    d. 115

30. In Jim's school, there are 3 girls for every 2 boys. There are 650 students in total. Using this information, how many students are girls?
    a. 260
    b. 130
    c. 65
    d. 390

31. A train traveling 50 miles per hour takes a trip lasting 3 hours. If a map has a scale of 1 inch per 10 miles, how many inches apart are the train's starting point and ending point on the map?
    a. 14
    b. 12
    c. 13
    d. 15

32. A traveler takes an hour to drive to a museum, spends 3 hours and 30 minutes there, and takes half an hour to drive home. What percentage of his or her time was spent driving?
    a. 15%
    b. 30%
    c. 40%
    d. 60%

33. At the store, Jan spends $90 on apples and oranges. Apples cost $1 each and oranges cost $2 each. If Jan buys the same number of apples as oranges, how many oranges did she buy?

34. If $3x = 6y = -2z = 24$, then what does $4xy + z$ equal?

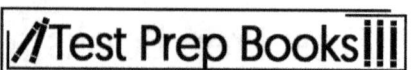

35. If $4x - 3 = 5$, then $x =$

36. What is the solution to $4 \times 7 + (25 - 21)^2 \div 2$?

37. Solve the following:

$$(\sqrt{36} \times \sqrt{16}) - 3^2$$

38. What is the overall median of Dwayne's current test scores: 78, 92, 83, 97?

39. What is the value of $x^2 - 2xy + 2y^2$ when $x = 2, y = 3$?

Answer Explanations for Practice Test #1

1. B: To simplify this inequality, subtract 3 from both sides to get $-\frac{1}{2}x \geq -1$. Then, multiply both sides by -2 (remembering this flips the direction of the inequality) to get $x \leq 2$.

2. D: This problem involves a composition function, where one function is plugged into the other function. In this case, the $f(x)$ function is plugged into the $g(x)$ function for each x-value. The composition equation becomes:

$$g(f(x)) = 2^3 - 3(2^2) - 2(2) + 6$$

Simplifying the equation gives the answer:

$$g(f(x)) = 8 - 3(4) - 2(2) + 6$$

$$8 - 12 - 4 + 6 = -2$$

3. C: An x-intercept is the point where the graph crosses the x-axis. At this point, the value of y is 0. To determine if an equation has an x-intercept of -2, substitute -2 for x, and calculate the value of y. If the value of -2 for x corresponds with a y-value of 0, then the equation has an x-intercept of -2. The only answer choice that produces this result is Choice C → $0 = (-2)2 + 5(-2) + 6$.

4. C: The conditional frequency of a girl being potty-trained is calculated by dividing the number of potty-trained girls by the total number of girls: $34 \div 52 = 0.65$. To determine the conditional probability, multiply the conditional frequency by 100: $0.65 \times 100 = 65\%$.

5. D: This system of equations involves one quadratic function and one linear function, as seen from the degree of each equation. One way to solve this is through substitution. Solving for y in the second equation yields $y = x + 2$. Plugging this equation in for the y of the quadratic equation yields

$$x^2 - 2x + x + 2 = 8$$

Simplifying the equation, it becomes $x^2 - x + 2 = 8$. Setting this equal to zero and factoring, it becomes:

$$x^2 - x - 6 = 0 = (x - 3)(x + 2)$$

Solving these two factors for x gives the zeros $x = 3, -2$. To find the y-value for the point, each number can be plugged in to either original equation. Solving each one for y yields the points $(3, 5)$ and $(-2, 0)$.

6. B: The slope will be given by $\frac{1-0}{2-0} = \frac{1}{2}$. The y-intercept will be 0, since it passes through the origin. Using slope-intercept form, the equation for this line is $y = \frac{1}{2}x$.

7. B: The table shows values that are increasing exponentially. The differences between the inputs are the same, while the differences in the outputs are changing by a factor of 2. The values in the table can be modeled by the equation $f(x) = 2^x$.

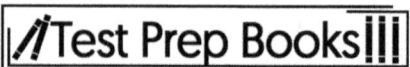

Answer Explanations for Practice Test #1

8. B: For the first card drawn, the probability of a King being pulled is $\frac{4}{52}$. Since this card isn't replaced, if a King is drawn first the probability of a King being drawn second is $\frac{3}{51}$. The probability of a King being drawn in both the first and second draw is the product of the two probabilities: $\frac{4}{52} \times \frac{3}{51} = \frac{12}{2652}$. This fraction, when divided by 12, equals $\frac{1}{221}$.

9. D: The expression is three times the sum of twice a number and 1, which is $3(2x + 1)$. Then, 6 is subtracted from this expression.

10. A: To expand a squared binomial, it's necessary use the *First, Inner, Outer, Last Method*.

$$(2x - 4y)^2$$

$$2x \times 2x + 2x(-4y) + (-4y)(2x) + (-4y)(-4y)$$

$$4x^2 - 8xy - 8xy + 16y^2$$

$$4x^2 - 16xy + 16y^2$$

11. B: The zeros of this function can be found by using the quadratic formula:

$$x = \frac{-b \pm \sqrt{b^2 - 4ac}}{2a}$$

Identifying *a*, *b*, and *c* can be done from the equation as well because it is in standard form. The formula becomes:

$$x = \frac{0 \pm \sqrt{0^2 - 4(1)(4)}}{2(1)} = \frac{\sqrt{-16}}{2}$$

Since there is a negative underneath the radical, the answer is a complex number: $x = \pm 2i$.

12. D: The expression is simplified by collecting like terms. Terms with the same variable and exponent are like terms, and their coefficients can be added.

13. D: The slope is given by the change in *y* divided by the change in *x*. Specifically, it's:

$$slope = \frac{y_2 - y_1}{x_2 - x_1}$$

The first point is (-5, -3), and the second point is (0, -1). Work from left to right when identifying coordinates. Thus, the point on the left is point 1 (-5,-3) and the point on the right is point 2 (0,-1).

Now we need to just plug those numbers into the equation:

$$slope = \frac{-1 - (-3)}{0 - (-5)}$$

Answer Explanations for Practice Test #1

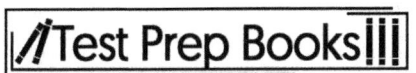

It can be simplified to:

$$slope = \frac{-1+3}{0+5}$$

$$slope = \frac{2}{5}$$

14. A: Finding the product means distributing one polynomial to the other so that each term in the first is multiplied by each term in the second. Then, like terms can be collected. Multiplying the factors yields the expression:

$$20x^3 + 4x^2 + 24x - 40x^2 - 8x - 48$$

Collecting like terms means adding the x^2 terms and adding the x terms. The final answer after simplifying the expression is:

$$20x^3 - 36x^2 + 16x - 48$$

15. B: The equation can be solved by factoring the numerator into $(x+6)(x-5)$. Since that same factor $(x-5)$ exists on top and bottom, that factor cancels. This leaves the equation $x + 6 = 11$. Solving the equation gives the answer $x = 5$. When this value is plugged into the equation, it yields a zero in the denominator of the fraction. Since this is undefined, there is no solution.

16. C: The common denominator here will be $4x$. Rewrite these fractions as:

$$\frac{3}{x} + \frac{5u}{2x} - \frac{u}{4}$$

$$\frac{12}{4x} + \frac{10u}{4x} - \frac{ux}{4x}$$

$$\frac{12x + 10u - ux}{4x}$$

17. B: There are two zeros for the given function. They are $x = 0, -2$. The zeros can be found a number of ways, but this particular equation can be factored into:

$$f(x) = x(x^2 + 4x + 4) = x(x+2)(x+2)$$

By setting each factor equal to zero and solving for x, there are two solutions. On a graph, these zeros can be seen where the line crosses the x-axis.

18. A: The equation is even because $f(-x) = f(x)$. Plugging in a negative value will result in the same answer as when plugging in the positive of that same value. The function:

$$f(-2) = \frac{1}{2}(-2)^4 + 2(-2)^2 - 6 = 8 + 8 - 6 = 10$$

yields the same value as:

$$f(2) = \frac{1}{2}(2)^4 + 2(2)^2 - 6 = 8 + 8 - 6 = 10$$

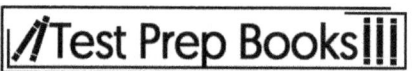

Answer Explanations for Practice Test #1

19. C: The expected value is equal to the total sum of each product of individual score and probability. There are 36 possible rolls. The probability of rolling a 2 is $\frac{1}{36}$. The probability of rolling a 3 is $\frac{2}{36}$. The probability of rolling a 4 is $\frac{3}{36}$. The probability of rolling a 5 is $\frac{4}{36}$. The probability of rolling a 6 is $\frac{5}{36}$. The probability of rolling a 7 is $\frac{6}{36}$. The probability of rolling an 8 is $\frac{5}{36}$. The probability of rolling a 9 is $\frac{4}{36}$. The probability of rolling a 10 is $\frac{3}{36}$. The probability of rolling an 11 is $\frac{2}{36}$. Finally, the probability of rolling a 12 is $\frac{1}{36}$.

Each possible outcome is multiplied by the probability of it occurring. Like this:

$$2 \times \frac{1}{36} = a$$

$$3 \times \frac{2}{36} = b$$

$$4 \times \frac{3}{36} = c$$

And so forth.

Then all of those results are added together:

$$a + b + c \ldots = expected\ value$$

In this case, it equals 7.

20. D: The addition rule is necessary to determine the probability because a 6 can be rolled on either roll of the die. The rule used is $P(A\ or\ B) = P(A) + P(B) - P(A\ and\ B)$. The probability of a 6 being individually rolled is $\frac{1}{6}$ and the probability of a 6 being rolled twice is $\frac{1}{6} \times \frac{1}{6} = \frac{1}{36}$. Therefore, the probability that a 6 is rolled at least once is $\frac{1}{6} + \frac{1}{6} - \frac{1}{36} = \frac{11}{36}$.

21. C: $y = 40x + 300$. In this scenario, the variables are the number of sales and Karen's weekly pay. The weekly pay depends on the number of sales. Therefore, weekly pay is the dependent variable (y), and the number of sales is the independent variable (x). Each pair of values from the table can be written as an ordered pair (x, y): (2,380), (7,580), (4,460), (8,620). The ordered pairs can be substituted into the equations to see which creates true statements (both sides equal) for each pair. Even if one ordered pair produces equal values for a given equation, the other three ordered pairs must be checked. The only equation which is true for all four ordered pairs is $y = 40x + 300$:

$$380 = 40(2) + 300 \rightarrow 380 = 380$$

$$580 = 40(7) + 300 \rightarrow 580 = 580$$

$$460 = 40(4) + 300 \rightarrow 460 = 460$$

$$620 = 40(8) + 300 \rightarrow 620 = 620$$

Answer Explanations for Practice Test #1

22. B: The car is traveling at a speed of five meters per second. On the interval from one to three seconds, the position changes by fifteen meters. By making this change in position over time into a rate, the speed becomes ten meters in two seconds or five meters in one second.

23. B: For an ordered pair to be a solution to a system of inequalities, it must make a true statement for BOTH inequalities when substituting its values for x and y. Substituting $(-3,-2)$ into the inequalities produces $(-2) > 2(-3) - 3 \rightarrow -2 > -9$ and $(-2) < -4(-3) + 8 \rightarrow -2 < 20$. Both are true statements.

24. D: The shape of the scatterplot is a parabola (U-shaped). This eliminates Choices A (a linear equation that produces a straight line) and C (an exponential equation that produces a smooth curve upward or downward). The value of *a* for a quadratic function in standard form ($y = ax^2 + bx + c$) indicates whether the parabola opens up (U-shaped) or opens down (upside-down U). A negative value for *a* produces a parabola that opens down; therefore, Choice B can also be eliminated.

25. D: Exponential functions can be written in the form: $y = a \cdot b^x$. The equation for an exponential function can be written given the y-intercept (*a*) and the growth rate (*b*). The y-intercept is the output (*y*) when the input (*x*) equals zero. It can be thought of as an "original value," or starting point. The value of *b* is the rate at which the original value increases ($b > 1$) or decreases ($b < 1$). In this scenario, the y-intercept, *a*, would be $1200, and the growth rate, *b*, would be 1.01 (100% of the original value + 1% interest = 101% = 1.01).

26. B: To simplify the given equation, the first step is to make all exponents positive by moving them to the opposite place in the fraction. This expression becomes $\frac{4b^3b^2}{a^1a^4} * \frac{3a}{b}$. Then the rules for exponents can be used to simplify. Multiplying the same bases means the exponents can be added. Dividing the same bases means the exponents are subtracted.

27. D: Dividing rational expressions follows the same rule as dividing fractions. The division is changed to multiplication, and the reciprocal is found in the second fraction. This turns the expression into $\frac{5x^3}{3x^2} * \frac{3y^9}{25}$. Multiplying across and simplifying, the final expression is $\frac{xy^8}{5}$.

28. C: The average is calculated by adding all six numbers, then dividing by 6. The first five numbers have a sum of 25. If the total divided by 6 is equal to 6, then the total itself must be 36. The sixth number must be $36 - 25 = 11$.

29. C: X = 150

Set up the initial equation.

$$\frac{2X}{5} - 1 = 59$$

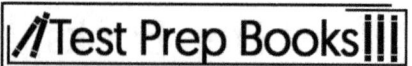

Add 1 to both sides.

$$\frac{2X}{5} - 1 + 1 = 59 + 1$$

Multiply both sides by 5/2.

$$\frac{2X}{5} \times \frac{5}{2} = 60 \times \frac{5}{2} = 150$$

$$X = 150$$

30. D: Three girls for every two boys can be expressed as a ratio: 3:2. This can be visualized as splitting the school into 5 groups: 3 girl groups and 2 boy groups. The number of students which are in each group can be found by dividing the total number of students by 5:

650 divided by 5 equals 1 part, or 130 students per group

To find the total number of girls, multiply the number of students per group (130) by the number of girl groups in the school (3). This equals 390, which is answer D.

31. D: First, the train's journey in the real world is 3 x 50 = 150 miles. On the map, 1 inch corresponds to 10 miles, so there is 150/10 = 15 inches on the map.

32. B: The total trip time is 1 + 3.5 + 0.5 = 5 hours. The total time driving is 1 + 0.5 = 1.5 hours. So, the fraction of time spent driving is 1.5/5 or 3/10. To get the percentage, convert this to a fraction out of 100. The numerator and denominator are multiplied by 10, with a result of 30/100. The percentage is the numerator in a fraction out of 100, so 30%.

33.

[Answer grid showing: 3 0]

One apple/orange pair costs $3 total. Therefore, Jan bought 90/3 = 30 total pairs, and hence, she bought 30 oranges.

Answer Explanations for Practice Test #1

34.

First solve for x, y, and z. So, $3x = 24, x = 8, 6y = 24, y = 4$, and $-2z = 24, z = -12$. This means the equation would be $4(8)(4) + (-12)$, which equals 116.

35.

Add 3 to both sides to get $4x = 8$. Then divide both sides by 4 to get $x = 2$.

36.

To solve this correctly, keep in mind the order of operations with the mnemonic PEMDAS (Please Excuse My Dear Aunt Sally). This stands for Parentheses, Exponents, Multiplication, Division, Addition, Subtraction. Taking it step by step, solve the parentheses first:

$$4 \times 7 + (4)^2 \div 2$$

Then, apply the exponent:

$$4 \times 7 + 16 \div 2$$

Multiplication and division are both performed next:

$$28 + 8 = 36$$

Answer Explanations for Practice Test #1

37.

Follow the order of operations in order to solve this problem. Solve the parentheses first, and then follow the remainder as usual.

$$(6 \times 4) - 9$$

This equals $24 - 9$ or 15.

38.

For an even number of total values, the median is calculated by finding the mean or average of the two middle values once all values have been arranged in ascending order from least to greatest. In this case, $(92 + 83) \div 2$ would equal the median 87.5.

39.

Start with the original equation: x- 2xy + 2y, then replace each instance of x with a 2, and each instance of y with a 3 to get:

$$2^2 - 2 \cdot 2 \cdot 3 + 2 \cdot 3^2 = 4 - 12 + 18 = 10$$

PSAT 8/9 Math Practice Test #2

1. If $6t + 4 = 16$, what is t?
    a. 1
    b. 2
    c. 3
    d. 4

2. The variable y is directly proportional to x. If $y = 3$ when $x = 5$, then what is y when $x = 20$?
    a. 10
    b. 12
    c. 14
    d. 16

3. A line passes through the point (1, 2) and crosses the y-axis at y = 1. Which of the following is an equation for this line?
    a. $y = 2x$
    b. $y = x + 1$
    c. $x + y = 1$
    d. $y = \frac{x}{2} - 2$

4. There are 4x + 1 treats in each party favor bag. If a total of 60x + 15 treats are distributed, how many bags are given out?
    a. 15
    b. 16
    c. 20
    d. 22

5. Apples cost $2 each, while oranges cost $3 each. Maria purchased 10 fruits in total and spent $22. How many apples did she buy?
    a. 5
    b. 6
    c. 7
    d. 8

6. What are the polynomial roots of $x^2 + x - 2$?
    a. 1 and -2
    b. -1 and 2
    c. 2 and -2
    d. 9 and 13

7. What is the y-intercept of $y = x^{5/3} + (x - 3)(x + 1)$?
    a. 3.5
    b. 7.6
    c. -3
    d. -15.1

75

8. $x^4 - 16$ can be simplified to which of the following?
   a. $(x^2 - 4)(x^2 + 4)$
   b. $(x^2 + 4)(x^2 + 4)$
   c. $(x^2 - 4)(x^2 - 4)$
   d. $(x^2 - 2)(x^2 + 4)$

9. $(4x^2y^4)^{\frac{3}{2}}$ can be simplified to which of the following?
   a. $8x^3y^6$
   b. $4x^{\frac{5}{2}}y$
   c. $4xy$
   d. $32x^{\frac{7}{2}}y^{\frac{11}{2}}$

10. If $\sqrt{1+x} = 4$, what is x?
    a. 10
    b. 15
    c. 20
    d. 25

11. Suppose $\frac{x+2}{x} = 2$. What is x?
    a. -1
    b. 0
    c. 2
    d. 4

12. A ball is thrown from the top of a high hill, so that the height of the ball as a function of time is $h(t) = -16t^2 + 4t + 6$, in feet. What is the maximum height of the ball in feet?
    a. 6
    b. 6.25
    c. 6.5
    d. 6.75

13. Five students take a test. The scores of the first four students are 80, 85, 75, and 60. If the median score is 80, which of the following could NOT be the score of the fifth student?
    a. 60
    b. 80
    c. 85
    d. 100

14. In an office, there are 50 workers. A total of 60% of the workers are women, and the chances of a woman wearing a skirt is 50%. If no men wear skirts, how many workers are wearing skirts?
    a. 12
    b. 15
    c. 16
    d. 20

15. A company invests $50,000 in a building where they can produce saws. If the cost of producing one saw is $40, then which function expresses the amount of money the company pays? The variable y is the money paid and x is the number of saws produced.
    a. $y = 50,000x + 40$
    b. $y + 40 = x - 50,000$
    c. $y = 40x - 50,000$
    d. $y = 40x + 50,000$

16. A six-sided die is rolled. What is the probability that the roll is 1 or 2?
    a. $\frac{1}{6}$
    b. $\frac{1}{4}$
    c. $\frac{1}{3}$
    d. $\frac{1}{2}$

17. A line passes through the origin and through the point (-3, 4). What is the slope of the line?
    a. $-\frac{4}{3}$
    b. $-\frac{3}{4}$
    c. $\frac{4}{3}$
    d. $\frac{3}{4}$

18.

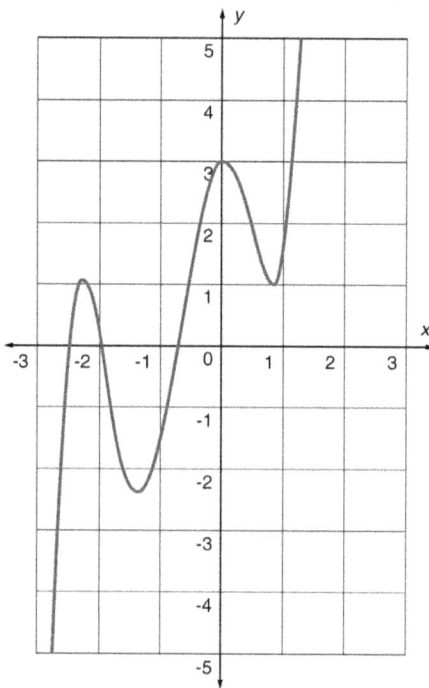

Which of the following functions represents the graph above?
a. $y = x^5 + 3.5x^4 - 6.5x^2 + 0.5x + 3$
b. $y = x^5 - 3.5x^4 + 6.5x^2 - 0.5x - 3$
c. $y = 5x^4 - 6.5x^2 + 0.5x + 3$
d. $y = -5x^4 - 6.5x^2 + 0.5x + 3$

19. Katie works at a clothing company and18. sold 192 shirts over the weekend. $1/3$ of the shirts that were sold were patterned, and the rest were solid. Which mathematical expression would calculate the number of solid shirts Katie sold over the weekend?

a. $192 \times \frac{1}{3}$

b. $192 \div \frac{1}{3}$

c. $192 \times (1 - \frac{1}{3})$

d. $192 \div 3$

20. What is the probability of randomly picking the winner and runner-up from a race of 4 horses and distinguishing which is the winner?

a. $\frac{1}{4}$

b. $\frac{1}{2}$

c. $\frac{1}{16}$

d. $\frac{1}{12}$

21. What is the next number in the following series: 1, 3, 6, 10, 15, 21, ... ?
    a. 26
    b. 27
    c. 28
    d. 29

22. How will the following algebraic expression be simplified: $(5x^2 - 3x + 4) - (2x^2 - 7)$?
    a. x^5
    b. $3x^2 - 3x + 11$
    c. $3x^2 - 3x - 3$
    d. $x - 3$

23. If $-3(x + 4) \geq x + 8$, what is the value of x?
    a. $x = 4$
    b. $x \geq 2$
    c. $x \geq -5$
    d. $x \leq -5$

24. For a group of 20 men, the median weight is 180 pounds and the range is 30 pounds. If each man gains 10 pounds, which of the following would be true?
    a. The median weight will increase, and the range will remain the same.
    b. The median weight and range will both remain the same.
    c. The median weight will stay the same, and the range will increase.
    d. The median weight and range will both increase.

25. If the ordered pair $(-3, -4)$ is reflected over the x-axis, what's the new ordered pair?
    a. $(-3, -4)$
    b. $(3, -4)$
    c. $(3, 4)$
    d. $(-3, 4)$

26. What's the midpoint of a line segment with endpoints $(-1, 2)$ and $(3, -6)$?
    a. $(1, 2)$
    b. $(1, 0)$
    c. $(-1, 2)$
    d. $(1, -2)$

27. A sample data set contains the following values: 1, 3, 5, 7. What's the standard deviation of the set?
    a. 2.58
    b. 4
    c. 6.23
    d. 1.1

No Calculator Questions

28. A solution needs 5 mL of saline for every 8 mL of medicine given. How much saline is needed for 45 mL of medicine?
    a. $\frac{225}{8}$ mL
    b. 72 mL
    c. 28 mL
    d. $\frac{45}{8}$ mL

29. A ball is drawn at random from a ball pit containing 8 red balls, 7 yellow balls, 6 green balls, and 5 purple balls. What's the probability that the ball drawn is yellow?
    a. $\frac{1}{26}$
    b. $\frac{19}{26}$
    c. $\frac{7}{26}$
    d. 1

30. If a car can travel 300 miles in 4 hours, how far can it go in an hour and a half?
    a. 100 miles
    b. 112.5 miles
    c. 135.5 miles
    d. 150 miles

31. How will the following number be written in standard form: $(1 \times 10^4) + (3 \times 10^3) + (7 \times 10^1) + (8 \times 10^0)$
    a. 137
    b. 13,078
    c. 1,378
    d. 8,731

32. What is the value of the sum of $\frac{1}{3}$ and $\frac{2}{5}$?
    a. $\frac{3}{8}$
    b. $\frac{11}{15}$
    c. $\frac{11}{30}$
    d. $\frac{4}{5}$

33. Ten students take a test. Five students get a 50. Four students get a 70. If the average score is 55, what was the last student's score?

34. $\frac{3}{25} =$

35. 6 is 30% of what number?

36. What is the value of the following expression?

$$\sqrt{8^2 + 6^2}$$

37. If Danny takes 48 minutes to walk 3 miles, how many minutes should it take him to walk 5 miles maintaining the same speed?

38. If $4x - 3 = 5$, then $x =$

39. If Sarah reads at an average rate of 21 pages in four nights, how many nights will it take her to read 140 pages?

Answer Explanations for Practice Test #2

1. B: First, subtract 4 from each side. This yields $6t = 12$. Now, divide both sides by 6 to obtain $t = 2$.

2. B: To be directly proportional means that $y = mx$. If x is changed from 5 to 20, the value of x is multiplied by 4. Applying the same rule to the y-value, also multiply the value of y by 4. Therefore, $y = 12$.

3. B: From the slope-intercept form, $y = mx + b$, it is known that b is the y-intercept, which is 1. Compute the slope as $\frac{2-1}{1-0} = 1$, so the equation should be $y = x + 1$.

4. A: Each bag contributes $4x + 1$ treats. The total treats will be in the form $4nx + n$ where n is the total number of bags. The total is in the form $60x + 15$, from which it is known $n = 15$.

5. D: Let a be the number of apples and o the number of oranges. Then, the total cost is $2a + 3o = 22$, while it also known that $a + o = 10$. Using the knowledge of systems of equations, cancel the o variables by multiplying the second equation by -3. This makes the equation $-3a - 3o = -30$. Adding this to the first equation, the b values cancel to get $-a = -8$, which simplifies to $a = 8$.

6. A: Finding the roots means finding the values of x when y is zero. The quadratic formula could be used, but in this case it is possible to factor by hand, since the numbers -1 and 2 add to 1 and multiply to -2. So, factor $x^2 + x - 2 = (x - 1)(x + 2) = 0$, then set each factor equal to zero. Solving for each value gives the values $x = 1$ and $x = -2$.

7. C: To find the y-intercept, substitute zero for x, which gives us:

$$y = 0^{5/3} + (0 - 3)(0 + 1) = 0 + (-3)(1) = -3$$

8. A: This has the form $t^2 - y^2$, with $t = x^2$ and $y = 4$. It's also known that $t^2 - y^2 = (t + y)(t - y)$, and substituting the values for t and y into the right-hand side gives $(x^2 - 4)(x^2 + 4)$.

9. A: Simplify this to:

$$(4x^2y^4)^{\frac{3}{2}} = 4^{\frac{3}{2}}(x^2)^{\frac{3}{2}}(y^4)^{\frac{3}{2}}$$

Now:

$$4^{\frac{3}{2}} = (\sqrt{4})^3 = 2^3 = 8$$

For the other, recall that the exponents must be multiplied, so this yields:

$$8x^{2 \cdot \frac{3}{2}} y^{4 \cdot \frac{3}{2}} = 8x^3 y^6$$

10. B: Start by squaring both sides to get $1 + x = 16$. Then subtract 1 from both sides to get $x = 15$.

11. C: Multiply both sides by x to get $x + 2 = 2x$, which simplifies to $-x = -2$, or $x = 2$.

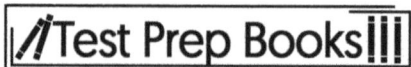

Answer Explanations for Practice Test #2

12. B: The independent variable's coordinate at the vertex of a parabola (which is the highest point, when the coefficient of the squared independent variable is negative) is given by $x = -\frac{b}{2a}$. Substitute and solve for x to get:

$$x = -\frac{4}{2(-16)} = \frac{1}{8}$$

Using this value of x, the maximum height of the ball (y), can be calculated. Substituting x into the equation yields:

$$h(t) = -16\frac{1}{8}^2 + 4\frac{1}{8} + 6 = 6.25$$

13. A: Lining up the given scores provides the following list: 60, 75, 80, 85, and one unknown. Because the median needs to be 80, it means 80 must be the middle data point out of these five. Therefore, the unknown data point must be the fourth or fifth data point, meaning it must be greater than or equal to 80. The only answer that fails to meet this condition is 60.

14. B: If 60% of 50 workers are women, then there are 30 women working in the office. If half of them are wearing skirts, then that means 15 women wear skirts. Since none of the men wear skirts, this means there are 15 people wearing skirts.

15. D: For manufacturing costs, there is a linear relationship between the cost to the company and the number produced, with a y-intercept given by the base cost of acquiring the means of production, and a slope given by the cost to produce one unit. In this case, that base cost is $50,000, while the cost per unit is $40. So, $y = 40x + 50,000$.

16. C: A die has an equal chance for each outcome. Since it has six sides, each outcome has a probability of $\frac{1}{6}$. The chance of a 1 or a 2 is therefore $\frac{1}{6} + \frac{1}{6} = \frac{1}{3}$.

17. A: The slope is given by:

$$m = \frac{y_2 - y_1}{x_2 - x_1} = \frac{0 - 4}{0 - (-3)} = -\frac{4}{3}$$

18. A: The graph contains four turning points (where the curve changes from rising to falling or vice versa). This indicates that the degree of the function (highest exponent for the variable) is 5, eliminating Choices C and D. The y-intercepts of the functions can be determined by substituting 0 for x and finding the value of y. The function for Choice A has a y-intercept of 3, and the function for Choice B has a y-intercept of -3. Therefore, Choice B is eliminated.

19. C: $\frac{1}{3}$ of the shirts sold were patterned. Therefore, $1 - \frac{1}{3} = \frac{2}{3}$ of the shirts sold were solid. Anytime "of" a quantity appears in a word problem, multiplication should be used. Therefore, $192 \times \frac{2}{3} = \frac{192 \times 2}{3} = \frac{384}{3} = 128$ solid shirts were sold. The entire expression is:

$$192 \times \left(1 - \frac{1}{3}\right)$$

Answer Explanations for Practice Test #2

20. D: $\frac{1}{12}$. The probability of picking the winner of the race is $\frac{1}{4}$ $\left(\frac{number\ of\ favorable\ outcomes}{number\ of\ total\ outcomes}\right)$. Assuming the winner was picked on the first selection, three horses remain from which to choose the runner-up (these are dependent events). Therefore, the probability of picking the runner-up is $\frac{1}{3}$. To determine the probability of multiple events, the probability of each event is multiplied: $\frac{1}{4} \times \frac{1}{3} = \frac{1}{12}$.

21. C: Each number in the sequence is adding one more than the difference between the previous two. For example, $10 - 6 = 4, 4 + 1 = 5$. Therefore, the next number after 10 is $10 + 5 = 15$. Going forward, $21 - 15 = 6, 6 + 1 = 7$. The next number is $21 + 7 = 28$. Therefore, the difference between numbers is the set of whole numbers starting at 2: 2, 3, 4, 5, 6, 7....

22. B: $3x^2 - 3x + 11$. By distributing the implied one in front of the first set of parentheses and the -1 in front of the second set of parentheses, the parenthesis can be eliminated:

$$1(5x^2 - 3x + 4) - 1(2x^2 - 7)$$

$$5x^2 - 3x + 4 - 2x^2 + 7$$

Next, like terms (same variables with same exponents) are combined by adding the coefficients and keeping the variables and their powers the same:

$$5x^2 - 3x + 4 - 2x^2 + 7$$

$$3x^2 - 3x + 11$$

23. D: $x \leq -5$. When solving a linear equation or inequality:

Distribution is performed if necessary: $-3(x + 4) \rightarrow -3x - 12 \geq x + 8$. This means that any like terms on the same side of the equation/inequality are combined.

The equation/inequality is manipulated to get the variable on one side. In this case, subtracting x from both sides produces $-4x - 12 \geq 8$.

The variable is isolated using inverse operations to undo addition/subtraction. Adding 12 to both sides produces $-4x \geq 20$.

The variable is isolated using inverse operations to undo multiplication/division. Remember if dividing by a negative number, the relationship of the inequality reverses, so the sign is flipped. In this case, dividing by -4 on both sides produces $x \leq -5$.

24. A: If each man gains 10 pounds, every original data point will increase by 10 pounds. Therefore, the man with the original median will still have the median value, but that value will increase by 10. The smallest value and largest value will also increase by 10 and, therefore, the difference between the two won't change. The range does not change in value and, thus, remains the same.

25. D: When an ordered pair is reflected over an axis, the sign of one of the coordinates must change. When it's reflected over the x-axis, the sign of the y coordinate must change. The x value remains the same. Therefore, the new ordered pair is $(-3, 4)$.

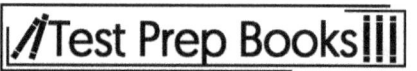

26. D: The midpoint formula should be used.

$$M = \left(\frac{x_1 + x_2}{2}, \frac{y_1 + y_2}{2}\right) = \left(\frac{-1+3}{2}, \frac{2+(-6)}{2}\right) = (1, -2)$$

27. A: First, the sample mean must be calculated. $\bar{x} = \frac{1}{4}(1 + 3 + 5 + 7) = 4$. The standard deviation of the data set is $\sigma = \sqrt{\frac{\Sigma(x-\bar{x})^2}{n-1}}$, and $n = 4$ represents the number of data points. Therefore:

$$\sigma = \sqrt{\frac{1}{3}[(1-4)^2 + (3-4)^2 + (5-4)^2 + (7-4)^2]}$$

$$\sqrt{\frac{1}{3}(9+1+1+9)} = 2.58$$

28. A: Every 8 ml of medicine requires 5 mL. The 45 mL first needs to be split into portions of 8 mL. This results in $\frac{45}{8}$ portions. Each portion requires 5 mL. Therefore:

$$\frac{45}{8} \times 5 = \frac{45*5}{8} = \frac{225}{8} \text{ mL is necessary}$$

29. C: The sample space is made up of $8 + 7 + 6 + 5 = 26$ balls. The probability of pulling each individual ball is $\frac{1}{26}$. Since there are 7 yellow balls, the probability of pulling a yellow ball is $\frac{7}{26}$.

30. B: 300 miles in 4 hours is $\frac{300}{4}$ = 75 miles per hour. In 1.5 hours, the car will go 1.5×75 miles, or 112.5 miles.

31. B: 13,078. The power of 10 by which a digit is multiplied corresponds with the number of zeros following the digit when expressing its value in standard form. Therefore:

$$(1 \times 10^4) + (3 \times 10^3) + (7 \times 10^1) + (8 \times 10^0)$$

$$10,000 + 3,000 + 70 + 8 = 13,078$$

32. B: $\frac{11}{15}$. Fractions must have like denominators to be added. The least common multiple of the denominators 3 and 5 is found. The LCM is 15, so both fractions should be changed to equivalent fractions with a denominator of 15. To determine the numerator of the new fraction, the old numerator is multiplied by the same number by which the old denominator is multiplied to obtain the new denominator. For the fraction $\frac{1}{3}$, 3 multiplied by 5 will produce 15. Therefore, the numerator is multiplied by 5 to produce the new numerator $\left(\frac{1 \times 5}{3 \times 5} = \frac{5}{15}\right)$. For the fraction $\frac{2}{5}$, multiplying both the numerator and denominator by 3 produces $\frac{6}{15}$. When fractions have like denominators, they are added by adding the numerators and keeping the denominator the same:

$$\frac{5}{15} + \frac{6}{15} = \frac{11}{15}$$

33.

Let the unknown score be x. The average will be $\frac{5\cdot50+4\cdot70+x}{10} = \frac{530+x}{10} = 55$. Multiply both sides by 10 to get $530 + x = 550$, or $x = 20$.

34.

The fraction is converted so that the denominator is 100 by multiplying the numerator and denominator by 4, to get $\frac{3}{25} = \frac{12}{100}$. Dividing a number by 100 just moves the decimal point two places to the left, with a result of 0.12.

35.

30% is $\frac{3}{10}$. The number itself must be $\frac{10}{3}$ of 6, or $\frac{10}{3} \times 6 = 10 \times 2 = 20$.

36.

8 squared is 64, and 6 squared is 36. These should be added together to get $64 + 36 = 100$. Then, the last step is to find the square root of 100 which is 10.

Answer Explanations for Practice Test #2

37.

Answer: 80

To solve the problem, a proportion is written consisting of ratios comparing distance and time. One way to set up the proportion is: $\frac{3}{48} = \frac{5}{x}$ $\left(\frac{distance}{time} = \frac{distance}{time}\right)$ where x represents the unknown value of time. To solve a proportion, the ratios are cross-multiplied:

$$(3)(x) = (5)(48) \rightarrow 3x = 240$$

The equation is solved by isolating the variable, or dividing by 3 on both sides, to produce $x = 80$.

38.

Answer: 2

Add 3 to both sides to get $4x = 8$. Then divide both sides by 4 to get $x = 2$.

39.

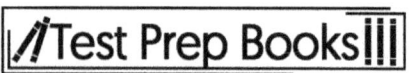

This problem can be solved by setting up a proportion involving the given information and the unknown value. The proportion is:

$$\frac{21\ pages}{4\ nights} = \frac{140\ pages}{x\ nights}$$

Solving the proportion by cross-multiplying, the equation becomes $21x = 4 * 140$, where $x = 26.67$. Since it is not an exact number of nights, the answer is rounded up to 27 nights. Twenty-six nights would not give Sarah enough time.

Dear PSAT 8/9 Test Taker,

We would like to start by thanking you for purchasing this study guide for the math section of your PSAT 8/9 exam. We hope that we exceeded your expectations.

Our goal in creating this study guide was to cover all of the topics that you will see on the math section of the test. We also strove to make our practice questions as similar as possible to what you will encounter on test day. With that being said, if you found something that you feel was not up to your standards, please send us an email and let us know.

We would also like to let you know about other books in our catalog that may interest you.

SAT

This can be found on Amazon: amazon.com/dp/1628458984

ACT

amazon.com/dp/1628458844

CLEP College Composition

amazon.com/dp/1628454199

AP Biology

amazon.com/dp/1628456221

We have study guides in a wide variety of fields. If the one you are looking for isn't listed above, then try searching for it on Amazon or send us an email.

Thanks Again and Happy Testing!
Product Development Team
info@studyguideteam.com

FREE Test Taking Tips DVD Offer

To help us better serve you, we have developed a Test Taking Tips DVD that we would like to give you for FREE. **This DVD covers world-class test taking tips that you can use to be even more successful when you are taking your test.**

All that we ask is that you email us your feedback about your study guide. Please let us know what you thought about it – whether that is good, bad or indifferent.

To get your **FREE Test Taking Tips DVD**, email freedvd@studyguideteam.com with "FREE DVD" in the subject line and the following information in the body of the email:

    a. The title of your study guide.

    b. Your product rating on a scale of 1-5, with 5 being the highest rating.

    c. Your feedback about the study guide. What did you think of it?

    d. Your full name and shipping address to send your free DVD.

If you have any questions or concerns, please don't hesitate to contact us at freedvd@studyguideteam.com.

Thanks again!

www.ingramcontent.com/pod-product-compliance
Lightning Source LLC
Chambersburg PA
CBHW081925170426
43200CB00014B/2838